Lecture Notes in Civil Engineering

Volume 315

Lecture Notes in Civil Engineering (LNCE) publishes the latest developments in Civil Engineering—quickly, informally and in top quality. Though original research reported in proceedings and post-proceedings represents the core of LNCE, edited volumes of exceptionally high quality and interest may also be considered for publication. Volumes published in LNCE embrace all aspects and subfields of, as well as new challenges in, Civil Engineering. Topics in the series include:

- Construction and Structural Mechanics
- Building Materials
- Concrete, Steel and Timber Structures
- Geotechnical Engineering
- Earthquake Engineering
- Coastal Engineering
- Ocean and Offshore Engineering; Ships and Floating Structures
- Hydraulics, Hydrology and Water Resources Engineering
- Environmental Engineering and Sustainability
- Structural Health and Monitoring
- Surveying and Geographical Information Systems
- Indoor Environments
- Transportation and Traffic
- Risk Analysis
- Safety and Security

To submit a proposal or request further information, please contact the appropriate Springer Editor:

- Pierpaolo Riva at pierpaolo.riva@springer.com (Europe and Americas);
- Swati Meherishi at swati.meherishi@springer.com (Asia—except China, Australia, and New Zealand);
- Wayne Hu at wayne.hu@springer.com (China).

All books in the series now indexed by Scopus and EI Compendex database!

Boominathan Adimoolam · Anirudhan I. V. ·
Sunil S. Basarkar · Amit Prashant
Editors

Deep Foundations for Infrastructure Development in India

Proceedings of DFI-India 2021 Annual Conference

Springer

Editors
Boominathan Adimoolam
Department of Civil Engineering
Indian Institute of Technology Madras
Chennai, Tamil Nadu, India

Sunil S. Basarkar
Department of Geotechnical Engineering
Afcons Infrastructure Ltd.
Mumbai, Maharashtra, India

Anirudhan I. V.
Geotechnical Solutions
Chennai, Tamil Nadu, India

Amit Prashant
Indian Institute of Technology Gandhinagar
Gandhinagar, Gujarat, India

ISSN 2366-2557 ISSN 2366-2565 (electronic)
Lecture Notes in Civil Engineering
ISBN 978-981-19-8597-3 ISBN 978-981-19-8598-0 (eBook)
https://doi.org/10.1007/978-981-19-8598-0

This Springer imprint is published by the registered company Springer Nature Singapore Pte Ltd.
The registered company address is: 152 Beach Road, #21-01/04 Gateway East, Singapore 189721, Singapore

Foreword

Deep Foundations Institute is an international non-profit association of engineers, contractors, manufacturers, equipment suppliers in the deep foundations and excavations operated from the United States of America. DFI brings together members through networking, education, communication and collaboration. DFI promotes the advancement of the deep foundation industry through technical committees, educational programmes, seminars, webinars, workshops and conferences, publication of guides and specifications, a peer-reviewed journal, a flagship magazine, research, government relations and outreach.

Deep Foundations Institute of India (DFI of India) was registered in Chennai with the Ministry of Company Affairs as a non-profit organisation and a regional chapter of DFI USA in 2013. DFI of India's mission is to support the Indian foundation industry on a continuous and sustained basis in measured and measurable steps, become very professional, and embrace new technologies for faster development of India. The chapter looks to provide a platform for continuous interaction with the Indian foundation industry stakeholders, including those international agencies with an India focus. Conferences, workshops, seminars and webinars are conducted regularly to highlight best practices and expertise worldwide. Interesting case studies and panel discussions are presented to raise technical awareness and propel advanced foundation design, construction, testing, contracting methods and risk-sharing in the Indian foundation industry.

Focussing on the key mission of technological enhancement of the deep foundation industry, DFI of India organises annual conferences with the common theme 'Deep Foundation Technologies for Infrastructure Development in India'. These yearly conferences showcase advanced deep foundation technologies and their application in Indian infrastructure projects through expert keynote presentations and contributory papers from industry and academia. The first conference was held in 2011 in Hyderabad, followed by the conferences at Chennai (2012), Mumbai (2013), New Delhi (2014), Bengaluru (2015), Kolkata (2016), Chennai (2017), Gandhinagar (2018), Hyderabad (2019) and two virtual conferences in 2020 and 2021.

The contributory papers are invited from the professional and academia through a call for abstracts. The abstracts are reviewed, and the authors of the selected abstracts are asked to develop the full-length paper. A pool of reviewers in the respective fields thoroughly examines the writings. At least two reviewers review and re-review each paper before accepting it for publication. Ninety-one abstracts were received for the conference in 2020 and 2021, and close to 50 full-length papers were submitted for review. After two rounds of reviews, 37 papers are selected for publication.

DFI of India considered it appropriate to publish the selected articles as a conference proceedings volume under the Springer series 'Lecture Notes in Civil Engineering'. Publishing the conference proceedings through a prestigious publisher, Springer, would enhance the visibility of the commendable works done in deep foundation technologies in India. DFI of India places on record its gratitude to Springer for bringing out this conference volume to benefit the geotechnical engineering community.

Chennai, India
2021

Mohan Ramanathan
DFI India 2020 and 2021 Conference Chair

Preface

Deep Foundations Institute of India conducted its annual conference on the main theme 'Deep Foundation Technologies for Infrastructure Development in India' DFI India 2021, virtually during 12–20 November 2021. Thirty-seven technical papers were presented during the conference. Each article was peer-reviewed and re-reviewed by two reviewers, each reviewer not reviewing more than six papers. Seventeen articles are included in this lecture series.

The papers are grouped under the sub-themes as follows:

1. Geotechnical Investigation, Testing, Instrumentation, Monitoring and Quality Management;
2. Ground Improvement Techniques;
3. Piling and Deep Foundation Techniques;
4. Earth Retention and Deep Excavation Support;
5. Research, Experimental and Numerical Methods in Deep Foundations and Deep Excavation Technologies; and
6. Safe and Efficient Geo-Construction.

Most of these articles were developed based on the field experiences requiring the theories put into practice. Advanced foundation testing procedures and new approaches for ground characterisation are discussed in four papers under the first topic. Two articles are featured under the theme of ground improvement techniques covering the effective use of stone columns, encased stone columns, sand compaction piles and grouting for weak rock.

There are three articles on piling and deep foundation techniques featuring new methods for constructing large diameter piles, soil–structure interaction and theoretical evaluation of deep foundation. Performance of deep excavation support systems and case studies are discussed in five papers under the theme Earth Retention and Deep Excavation Support.

There are three interesting articles under themes five and six.

We, the editors, sincerely hope that the articles covered in this volume would immensely help the professionals and academicians alike to improve the work practice.

Chennai, India — Boominathan Adimoolam
Chennai, India — Anirudhan I. V.
Mumbai, India — Sunil S. Basarkar
Gandhinagar, India — Amit Prashant

Contents

About the Editors

Dr. Boominathan Adimoolam is an active academician, researcher and consultant in the area of geotechnical engineering. His expertise includes soil dynamics and earthquake geotechnical engineering, deep foundations and ground improvement. He received his Ph.D. in Geotechnical Engineering from Moscow Civil Engineering Institute in 1986.

Dr. A. Boominathan was a Professor of Geotechnical Engineering at the Department of Civil Engineering at IIT Madras until his retirement in June 2021. He has guided 16 Ph.D., 11 M.S. by research and several M.Tech. students. He has authored more than 196 scientific publications, including more than 62 articles in refereed journals.

Dr. A. Boominathan is a member of the editorial board of the *Journal of Earthquake & Tsunami*. He is a fellow of the Indian Geotechnical Society, a National Executive member of Deep Foundation Institute (DFI) of India and Associate Member of American Society of Civil Engineers (ASCE). He is also an honorary fellow of the National Engineering Academy of the Republic of Kazakhstan. He is a member of several technical committees of the International Society of Soil Mechanics and Geotechnical Engineering (ISSMGE). He has received the IGS Shri M. S. Jain Biennial Prize in 2004, and the IGS Prof. C. S. Desai Biannual Prize in 2012. Recently he has been selected for the A. S. Arya Disaster Prevention Award, a research award of IIT Roorkee.

Anirudhan I. V. is a Geotechnical Consultant based in Chennai, India. His career as an independent consultant started in 1987 after associating with a leading consultancy firm for seven years. His major field of expertise are in deep foundations, ground improvement and geotechnical characterisation. His organisation, Geotechnical Solutions, executed more than 2000 small and medium geotechnical investigations and provided foundation recommendations. He graduated in Civil Engineering with a specialisation in Geotechnical Engineering from IIT Kanpur, India, in 1981 after obtaining B.Sc. Engg. (Hons) in Civil Engineering from the University of Calicut in the year 1979.

He is actively involved in the activities of the Indian Geotechnical Society and Deep Foundations Institute of India, and he has conducted several conferences, workshops and seminars. He was a member of TC-302 of ISSMGE on Forensic Engineering, TC-301 of ISSMGE on Historic Sites and the convener of TC-04 of the Indian Geotechnical Society on Geotechnical Investigation. He has authored and published several technical papers, the guideline manuals on lateritic soils, the guideline manual on geotechnical investigation, the design manual for continuous flight augured piles. He is currently the chairman of Deep Foundations Institute of India, and he also served as the chairman and honorary secretary of the Chennai Chapter of the IGS.

Dr. Sunil S. Basarkar is the Head of Geotechnical Engineering Department at Afcons Infrastructure Ltd Mumbai.

He has a total experience of about 18 years in the field and 16 years association with Academia in area of geotechnical engineering. Dr. Basarkar has published more than 50 papers in various national and international conferences and reputed journals. He is a Core Committee Member of DFI of India. Dr. Basarkar is also a fellow of the Institution of Engineers (India), Indian Geotechnical Society, and Institute of Bridge Engineering. Dr. Basarkar also serves as an Editor of DFI Journal (USA).

Dr. Amit Prashant is Professor of Civil Engineering and Dean of Research and Development at IIT Gandhinagar. He is currently the Chairman of the Ahmedabad chapter of Indian Geotechnical Society and also a Member of the Board of Governors of SVNIT Surat. He completed his Bachelors of Engineering degree from IIT Roorkee in 1997. After two years of consulting experience, he went for higher studies at Clarkson University, NY and then to University of Tennessee, Knoxville. After completing his doctoral studies and one year of post-doctoral research, he worked at IIT Kanpur as Assistant Professor during 2005–2010, and then moved to IIT Gandhinagar in 2010. He has been actively involved in academic administration, as Chairman, SUGC at IIT Kanpur earlier and for six years as Dean, Academic Affairs at IIT Gandhinagar. With that experience, he was also appointed as the Officiating Director of IIT Gandhinagar during July-September 2019. His research interests are in constitutive modelling of soils, earthquake geotechnical engineering and applications of geosynthetics, which including experimental investigations and numerical modelling. He has done field projects related to hill slopes, excavations, canals, ash-dyke, bridge foundations and barrage, etc. He has conducted several industry courses in earthquake engineering, FEM and geotechnical investigations, which have been quite popular. He also coordinates faculty development programs for last six years under the aegis of TEQIP Scheme of the Government of India for engineering colleges in the country.

Geotechnical Investigation, Testing, Instrumentation, Monitoring and Quality Management

Advanced Geotechnical Investigation and Data Interpretation for Complex Underground Structures

M. Kalaiselvi, Y. Kaushik Sarma, and G. D. Raju

Abstract In recent times, development of underground space for variety of purposes is becoming more viable alternative to surface facilities with an ultimate aim of conserving surface space for various reasons and for strategic purposes. However, construction of underground structures continues to be expensive, with project cost rising rapidly and often significantly exceeding the preconstruction estimate. One of the most desirable and optimistic ways to improve the economic feasibility of underground construction is adopting a proper and systematic geotechnical investigation process. As variation in geology plays a dominant role in cost, feasibility, safety, and behavior of underground structure, serious consideration has to be given for geotechnical investigation. As soon as the investigation data is obtained, it should be properly interpreted and evaluated to get a meaningful outcome for greater savings in cost and time through optimum design. The basic objective of this paper is to recommend guidelines for planning and executing an effective multiphase geotechnical investigation program with advanced methods depending on the complexity of project and timely interpretation of the investigation data. This paper highly emphasizes on deriving the stratigraphy, structural features, presence of water pockets, in situ stress condition, and strength parameters through field and laboratory investigations. From this study, geotechnical engineers are expected to achieve a great extent of understanding in interpretation of investigation data and collaborating interrelated geotechnical information to derive the key aforementioned inputs. As a result, this will help in designing adequate support system for the underground structures, eradicating the possibilities of overdesign.

Keywords Geotechnical investigation · Guidelines · Field testing · Geotechnical parameters · Underground structure

M. Kalaiselvi (✉) · Y. K. Sarma · G. D. Raju
L&T Construction, Chennai, Tamil Nadu, India
e-mail: kalai.geotech@gmail.com

G. D. Raju
e-mail: gdsraju@lntecc.com

B. Adimoolam et al. (eds.), *Deep Foundations for Infrastructure Development in India*, Lecture Notes in Civil Engineering 315, https://doi.org/10.1007/978-981-19-8598-0_1

1 Introduction

Geotechnical investigation for complex underground structures, such as tunnels with intersections, caverns, slopes, etc., where conducting numerous geotechnical investigations and correlating their results become a crucial aspect of designing a desirable and economically feasible underground structure. Correlating the results of commonly practised geotechnical investigation methods with some recently developed techniques brings out a clear and reliable information of the project area. The scope of this paper is to bring clarity of investigation planning and interpreting the gathered data through various investigations to get relevant geotechnical input for the design of underground structures. The proposal for site investigation planning is mainly focussed on rocky strata, as there is limited availability of information and guidelines for investigation of rockmass in which the structure is to be excavated. It is highly important to propose a systematic site investigation plan for a complex underground projects with regions of significant faulting, shearing, alteration, and folding. Planning of site investigation should include selection of applicable investigation methods based on the type of structure and proposal of intervals between investigation methods such that it covers the extent of project area including critical locations along the structure. The aforementioned knowledge can be obtained by studying the project site and its surrounding area through reconnaissance survey. The investigation should be planned in such a way that the complete data is properly utilized and providing additional information at right time. The comprehensive site investigation comprising surface mapping, geophysical surveys, test borings in soil and rock, oriented core readings, and in situ and laboratory testing and the results shall be precisely interpreted to provide a geological model.

A multiphase site investigation program is proposed post preliminary investigations as shown in Table 1 and it is largely divided into two phases, namely, (i) confirmatory geotechnical investigation before excavation (during design phase) and (ii) confirmatory geotechnical investigation during excavation.

On completion of site investigation, an equal importance shall be given to interpretation of gathered data through various investigations. In order to get meaningful design input from raw site investigation results, it is emphasized to carefully read and interpret the data with relative engineering judgement. A complete picture of the site geology can be achieved only if the geological data and interpretations from surface mapping, borings, and drift tunnels are combined and well correlated.

Further sections explain the two confirmatory geotechnical investigations and its data interpretation.

Table 1 Multiphase geotechnical investigation program

S. no	Investigation		Purpose
1	Confirmatory geotechnical investigation before construction	Geological surface mapping	Contouring
2		Seismic refraction test	Stratigraphy
3		Borehole investigation	Stratigraphy and structural geology
4		Seismic cross-hole tomography	
5		Core drilling	Physical and strength properties
6		Borehole televiewer	Structural geology
7	Confirmatory geotechnical investigation during construction	Geological 3D face mapping	Confirmation of geology and parameters
8		Probe hole drilling	
9		Drift investigation	

2 Confirmatory Geotechnical Investigation Before Excavation

(a) ***Geological surface mapping***

Entry level geological investigations include detailed mapping of the exposed rock and outcrop at the project site, which needs to be performed by an experienced geologist. Output of this investigation is expected to provide site-specific data on nature of joints, degree of weathering, structural features, and other essential information on rock quality and structure.

The mapping campaign includes but shall not be limited to the following information:

(i) Geologic mapping shall characterize and document the condition of rockmass or outcrop for its classification. The mapping symbols or legend shall be in accordance with IS 7422-Part 5 and wherever appropriate mapping symbols are not available; they shall be referred from ASTM D4879.

(ii) Details and reporting of mapping shall be done as per standards (ASTM D4879, for instance). The mapping details as a minimum shall include discontinuity mapping, rockmass response mapping, rock descriptions, rock defects, rock structure and rock condition, fluid descriptions, photographic record, technical discussion of geotechnical maps, and potential problems description.

(iii) International Society for Rock Mechanics (ISRM 1981) and/or IS 11315 Part 1 to 10 shall be referred for describing discontinuities.

(iv) Mapping shall capture and document all surface features like slide potential near portal and shaft locations; faults, degree of rock weathering, groundwater springs, volcanic history, presence of sensitive minerals, stress relief cracks, zones of deep weathering or presence of talus or boulders.

(v) Joints, faults, and bedding planes shall be mapped, and their orientations are plotted by stereographic projection. This data shall then be projected to the elevation of the underground structure location to assess geological conditions.
(vi) Nearby natural exposures and man-made exposures (quarries and cuttings) at the site shall be considered to completely characterize geological material likely to be encountered.
(vii) Extrapolation of data shall be avoided, and interpolation shall be made with professional discretion.
(viii) As the construction work proceeds, fresh geological information on various excavations shall be logged and previous records updated.

(b) ***Geophysical methods***

Two geophysical tests, namely, seismic cross-hole tomography test and seismic refraction survey shall be conducted as per IS 13372-part 2 and IS 15681, respectively.

Seismic cross-hole tomography produces a 3D interpretation of the subsurface. It involves the measurement of travel times of seismic ray paths between two or more boreholes in order to derive an image of seismic velocity in the intervening ground. The level of detail depends upon the distance between holes, the power of the source, and the properties of the rock or soil mass. The method can be used for both indurated and nonindurated geomaterials. The test data shall be utilized to arrive at dynamic and elastic parameters, and also to identify subsurface and fault zones through compression and shear wave velocities.

Requisite components in the cross-hole tomography method for accurate field measurement of shear wave and compression wave velocities and their attenuation will include mechanical sources which are strong, data acquisition system (an amplifier, signal enhancement equipment, timing device, recorder, and a time break unit), and a seismic receiver as per IS 13372-part 2.

Seismic refraction survey is the quickest and often most economical method of obtaining general information over large land areas, or in areas with difficult access, such as mountainous regions or large water bodies. They are particularly useful in investigating shallow rock conditions.

This test shall be executed on the surface as identified during geological mapping. It is recommended to conduct seismic refraction tests along the tunnel axis covering the entire length of tunnel.

(c) ***Borehole drilling***

Boreholes are carried out to obtain information about the sub-strata profile, its nature and strength, and to collect soil or rock samples for strata identification and conducting laboratory tests. Boring will be carried out in accordance with the provisions of IS 1892.

In situ tests such as hydrofracturing test, water permeability test, and dilatometer test shall be conducted and undisturbed samples can be obtained in the boreholes.

Representative disturbed samples will be preserved for conducting various tests in the laboratory. Water table in the borehole shall be carefully recorded and reported. SPT N test shall be conducted in overburden as per IS 2131.

On completion of the boreholes, it should be backfilled if not specifically required for another purpose; any hole will be filled with cement slurry. Colored cement will be used in order to make easier the recognition of investigation holes in underground works.

(d) ***Borehole televiewer***

Borehole cameras (TV and photography) furnish images of borehole walls in rock-masses. It complements the structural assessment of the rockmass by providing digital images and quantified data of discontinuities. These can be conducted in vertical and inclined exploratory boreholes. The images of the hole shall be provided with visual highlighting of main discontinuities and calculated altitude (azimuth/dip angle or dip angle/dip direction). Description of various types of discontinuity filling (e.g., quartz, calcite, and clay) can also be recorded.

One or combination of following borehole logging methods shall be used for selected direct exploratory boreholes.

1. Borehole camera: optical systems using visible wavelength and providing true color images; to be used in holes without water filling or clear water.
2. Borehole televiewer using an acoustic pulse (ultrasonic wavelength) and providing false color images; to be used only in boreholes filled with water.

3 Confirmatory Geotechnical Investigation During Construction

(a) ***Geological 3D face mapping***

3D face, in general, shall include the following:

(i) Tunnel mapping

The exposed face conditions shall be documented in cross-sectional sketches (face mapping) and drawn at frequent intervals as the tunnel advances.

In addition, following information shall be included in the face mapping but not limited to

- station location for the cross section;
- date and time the face mapping was prepared;
- name of the individual preparing the face map;
- material classification;
- location of interface boundaries between these materials;
- rock jointing including orientation of principal joints and joint descriptions;
- shear zones;
- seepage conditions and their approximate locations on the face;

- Observed ground behavior with reference to likely instability or squeezing material at the face.

(ii) Impediments and Obstructions
Likelihood of impediments during tunneling like presence of boulders, foundation piles, grouted or cemented material, and any other significant observations.

Sketches presenting similar information where the perimeter rock is left exposed shall be prepared for the tunnel walls and roof. The face maps shall be used to accurately document conditions exposed during tunneling, and to develop a detailed profile of subsurface conditions along the tunnel horizon.

(b) ***Probe holes***

Probing ahead of the tunnel face shall be carried out to determine general ground conditions in advance of excavation, and to identify and relieve water pressures in any localized zones of water-bearing soils or rock joints. Probing also provides an early indication of the type of ground supports that may be needed as the excavation progresses.

Probing typically consists of drilling horizontal to sub-horizontal holes from the tunnel excavation face by percussion drilling or rotary drilling methods.

(c) ***Drift investigation***

The geological mapping in drift shall be conducted and prepared in accordance with ASTM D4879 together with the following requirements:

- Apparatus to be used shall conform to ASTM D4879 as a minimum.
- The mapping shall be performed by an experienced geologist. He shall become familiar with the regional and local geology prior to mapping. He shall supervise and be responsible for the mapping.
- The objective of the mapping is to develop an appropriate visual and quantitative representation of the rockmass and associated features, such as excavation supports, in situ testing locations, and groundwater inflows. The details of this representation will vary with the rock type, the size of the opening, and the intended use of the data.
- Time of mapping, relative to time of excavation, is influenced by time-dependent rock behavior (namely, swelling, slaking, stress relaxation, etc.), but mapping shall be performed as soon as exposures are made.
- The mapping shall be accomplished after excavation; however, the mapping shall be completed before construction activity modifies or obscures the surface condition.
- Mapping shall be done on all excavated surfaces including the floor of the opening.
- Full-periphery mapping technique shall be done as explained in ASTM D4879 and wherever details are not available in former, they shall be referred to USACE EM 1100-1-1804. This representation shall show all exposed features, including seepage.

- Preparation:
  Surfaces of the excavation shall be cleaned by water or air immediately prior to mapping; consideration to the potentially detrimental effects of cleaning on the natural materials shall be given.
  Stationing or survey control points shall be marked on the walls at appropriate intervals immediately prior to mapping.
- The mapping symbols or legend shall be in accordance with IS 7422-Part 5 and wherever the details are not available in former, shall be in accordance with ASTM D4879.
- Details and reporting of mapping shall be done in accordance with ASTM D4879. The mapping details as a minimum shall include discontinuity mapping, rockmass response mapping, rock descriptions, rock defects, rock structure and rock condition, fluid descriptions, full periphery mapping, photographic record, technical discussion of geotechnical maps, and potential problems description.

4 Geological and Geotechnical Interpretation

The investigation data obtained during the design and excavation stage is to be judgmentally interpreted to get quantitative design parameters in a qualitative manner. Geotechnical investigation methods such as surface mapping, seismic refraction methods, and borehole logging shall be performed during the design stage to obtain quantitative parameters such as joint orientation, joint parameters, rockmass type, in situ stress ratio, and rockmass shear parameters. Table 2 presents the different quantitative and qualitative parameters that can be derived through various investigations.

4.1 Development of Stratigraphy

Stratigraphy of the underground geology shall be developed from the results of seismic refraction test, borehole investigation, and borehole tomography. From seismic refraction test, stratum depths based on velocities shall be obtained. Geologic sections can be interpreted based on depths of changes in strata. Material types are judged from correlations with velocities. Table 3 relates subsurface strata with seismic velocity. Based on this measure, the subsurface strata shall be bifurcated into number of layers.

Table 2 Geological and geotechnical investigation and interpretation

Investigation	Quantitative parameters	Qualitative parameters (direct)
Surface mapping	.• Geological fractures like folds, faults, lineaments, etc. • Surficial joint parameters • Type of rockmass • Any probable anomaly	• Joint sets • Aperture size • Size and orientation of any major geological feature • Geological mapping
Core logging of boreholes	• Stratigraphy • Quality of rockmass • Core loss and water loss • Weathering of rockmass	• RQD and core recovery • Preliminary joint details • Joint alterations and joint infillings
Seismic refraction test	• Stratigraphy • Near to surface geological features like progression of faults and lineaments	• Density of rockmass • Depth of weathering zone • Stratigraphy up to a depth of 60–70 m
Cross-hole tomography	• Stratigraphy • Deeper depth geological features like progression of faults and lineaments	• Density of rockmass • Depth of weathering zone • Stratigraphy up to a depth of 200 m for defined zone
Borehole televiewer	• Confirmation of joints and other features of core losses • Shear seams • Joint details	• Joint numbers • orientation • Types of joints (open or closed)
Hydraulic fracturing test	• In situ stress ratio	• Orientation and magnitude of major and minor principal stresses
Water permeability test (Lugeon test)	• Permeability of rockmass	• Rockmass type
Dilatometer test	• Deformation modulus of rockmass	• Deformation properties
Laboratory testing	• Type of rockmass • Mineral composition	• Unconfined compressive strength • Tensile strength • Cohesion • Friction angle

4.2 Identification of Prominent Geological Features

From surface geological mapping, borehole televiewer logging, and face mapping data, orientation of discontinuities such as joints, foliations, shear zones, and fracture zones is obtained. When there is a huge amount of joint set data, DIPS software can be used to identify prominent joint sets with its orientation. DIPS is a stereographic projection program with stereonets and histograms showing poles of all the projected joint set data through input. This program can also provide joint parameters such as Jn, Jr, and Ja depicting the joint set, roughness characteristics, and joint infillings corresponding to the joint set cluster.

Table 3 Engineering classification of rockmass based on seismic velocity (Bell 1992)

Class	Subsurface strata type	Seismic velocity (P wave) (m/s)	Likely rock types/strata
I	Overburden	Up to 1500	Loose and compact soil, fine and coarse grain soil/sand/clay, rock fragments, slope debris, etc.
II	Weathered rock	1500 – 2500	Highly fractured and fragmented rock, large rock boulders
III	Jointed rockmass	2500 – 3500	Rocks with closely spaced and intersecting joints. Can be strengthened by consolidation grouting
IV	Hard rock	3500 – 4500	Hard rock with in situ strength in the range of 80–150 MPa. Good stand-up time upon excavation
V	Massive rock	>4500	Hard rock with intact sample strength of over 100 MPa. Hard and massive granites/limestone or similar type of rock types

The identified discontinuities shall be overlapped with the alignment of any underground structure for structurally controlled stability analysis. This helps in identification of potential wedges such as sliding wedges and falling wedges which are formed in underground structures. When a free face created by excavation intersects structural features, such as bedding planes and joints, the rockmass separates into discrete but interlocked pieces. This process of identifying major joint sets assists in finalizing the alignment of a structure to its favorable orientation with respect to joint sets and designing appropriate support system to avoid the anticipated wedge failures beforehand. The orientations of many types of excavations can be, and normally are, adjusted to avoid the maximum effect of unfavorably oriented major joints.

4.3 Groundwater Condition

There are many factors that influence the design of underground structure, in which the existence of ground water is the most important factor. Under high groundwater table condition, the seepage situation becomes more complex and it is more difficult to control the leakage of groundwater flow into a tunnel. To avoid any catastrophic failures of any underground structure such as slope failure and huge seepage inrush, it is mandatory to identify the location of groundwater level or perched water table if any.

The groundwater level shall be assessed by using dip meter in boreholes at various intervals. These levels can be later plotted into a graph to get the groundwater table along the alignment of any structure. As this ground water level is identified during the design stage, it shall be confirmed through drift investigation in case of large cavern excavation. In addition, probe holes shall be drilled ahead of the excavation face. Data such as drilling rates, water flow, drilling fluid behavior, etc. from probe holes be used to give an indication of the rockmass quality. However, this information has to be interpreted by an Engineering Geologist.

4.4 In Situ Stress Conditions

In situ stress magnitudes and orientations play a very important role in geological engineering. When an opening is excavated in rock, in situ stress field is locally disrupted, and a new set of stresses are induced in the rock surrounding the opening. Knowledge of the magnitudes and directions of these in situ and induced stresses is an essential component of underground excavation design. This is true because in many cases, as the induced stress exceeds the strength of the rock, it may result in instability with serious consequences due to excavations and so it must be adequately analyzed before any tunneling or underground excavation works.

In a majority of projects, the major principal stress is vertical, i.e., the weight of the overlying material. However, it has been found from measurements made throughout the world that horizontal stresses in the near-surface vicinity, defined as 30 m or less, can be one and one-half to three times higher than the vertical stress. Recognition of this condition during the design phase of investigations is very important. Where high horizontal stresses occur at a project site, the stability of cut slopes and tunnel excavations is affected.

The direction of principal stress can be evaluated by observing the databases on geological, topographical, and tectonic information in the project area. Borehole breakouts can also be observed through borehole televiewer and this gives an idea of the direction of minimum principal stress in breakout direction due to stress orientation with respect to rock strength.

However, in situ testing is more reliable method for obtaining the magnitude and direction of stresses. The three most common methods for determining in situ stresses are the over-coring, hydro-fracture, and flat jack techniques. These test procedures shall be confirmed to IS 13946 (Part 1, 2, 3, and 4).

The minimum and maximum in situ stress ratios ($k_{\min}$ and $k_{\max}$) shall be used in deciding the alignment and dimension of any underground structure to have favorable stress orientation around the tunnel opening.

4.5 *Strata Characterization*

(a) ***Q system***

Barton et al. (1974) of the Norwegian Geotechnical Institute (NGI) proposed a Tunneling Quality Index (Q) for the determination of rockmass characteristics. Direct assessment of rockmass quality can be evaluated using Q-classification system and the guidelines for classification using Q-logging as given in Barton and Grimstad (2014) shall be followed. The numerical value of Q varies on a logarithmic scale from 0.001 to a maximum of 1,000 and is defined by

$$Q = \frac{\text{RQD}}{J_n} \times \frac{J_r}{J_a} \times \frac{J_w}{\text{SRF}} \tag{1}$$

where RQD = rock quality designation, J_n = joint set number, J_r = joint roughness number, J_a = joint alteration number, J_w = joint water reduction factor, and SRF = stress reduction factor.

The first quotient (RQD/J_n), representing the structure of the rockmass, is a crude measure of the block or particle size. The second quotient (J_r/J_a) represents the roughness and frictional characteristics of the joint walls or filling materials. The third quotient (J_w/SRF) consists of stress parameters.

Q-value can be assessed based on the above six parameters from face mapping and borehole data. Histograms of Q-parameters based on Q-logging can be made for the ease of understanding the rockmass quality in the structure zone.

(b) ***Geological strength index***

The Geological Strength Index (GSI) was introduced by Hoek (1994). Hoek, Kaiser, and Bawden (1995) provide a number which, when combined with the intact rock properties, can be used for estimating the reduction in rockmass strength for different geological conditions. GSI is estimated based on rock structure type and joint surface conditions. In order to use the Hoek–Brown criterion for estimating the strength and deformability of jointed rock masses, three "properties" of the rock mass have to be estimated. These are uniaxial compressive strength of the intact rock pieces, value of the Hoek–Brown constant (m_i) for these intact rock pieces, and value of the Geological Strength Index (GSI) for the rock mass. These GSI values can be correlated with Q for various rock classes.

4.6 *Derivation of Strength Parameters*

For any underground design, derivation of strength parameters of rockmass is one of the significant steps of design methodology. It is challenging to get rockmass parameters directly through lab test results. Design strength parameters such as cohesion, friction angle, and deformation modulus of rockmass can be calculated based on rockmass characterization (Q and GSI) using following correlations:

$$\sigma_{cm} \approx 5\gamma {Q_c}^{1/3} \tag{2}$$

$$c \approx \frac{\text{RQD}}{J_n} \times \frac{1}{\text{SRF}} \times \frac{\sigma_c}{100} \tag{3}$$

$$\varphi = \tan^{-1}\left(\frac{J_r}{J_a} \times \frac{J_w}{1}\right) \tag{4}$$

$$E_{rm} \approx 10 \times {Q_c}^{1/3} \tag{5}$$

where σ_{cm} = rockmass compressive strength, c = cohesion of rockmass, φ = friction angle of rockmass, and E_{rm} = rockmass modulus.

Cohesion and friction angle of rockmass are estimated by taking into account Hoek–Brown parameters and joint condition. The method derived by Hoek and Marinos (2007) or Roclab is used to determine the cohesion and friction angle of rockmass and detailed below.

UCS value of intact rock, GSI, material constant (m_i), and disturbance factor are required for the estimation of rockmass properties using the following equations (Hoek and Brown 1997):

$$m_b = m_i \exp\left(\frac{\text{GSI} - 100}{28 - 14D}\right) \tag{6}$$

$$s = \exp\left(\frac{\text{GSI} - 100}{9 - 3D}\right) \tag{7}$$

$$a = \frac{1}{2} + \frac{1}{6}\left(e^{-\text{GSI}/15} - e^{-20/3}\right) \tag{8}$$

$$\sigma_{3n} = \frac{{\sigma'}_{3\max}}{\sigma_{ci}} \tag{9}$$

$$c' = \frac{\sigma_{ci}\left[(1+2a)s + (1-a)m_b{\sigma'}_{3n}\right](s + m_b{\sigma'}_{3n})^{a-1}}{(1+a)(2+a)\sqrt{1 + \left(6am_b(s + m_b{\sigma'}_{3n})^{a-1}\right)/((1+a)(2+a))}} \tag{10}$$

$$\varphi' = \sin^{-1}\left[\frac{6am_b(s + m_b{\sigma'}_{3n})^{a-1}}{2(1+a)(2+a) + 6am_b(s + m_b{\sigma'}_{3n})^{a-1}}\right] \tag{11}$$

$$E_{rm}(\text{GPa}) = \left(1 - \frac{D}{2}\right)\sqrt{\frac{\sigma_{ci}}{100}} \times 10^{(\text{GSI}-10)/40} \tag{12}$$

where σ_{3n} = ratio of confining stress to intact compressive strength, σ'_{3max} = confining stress, σ_{ci} = intact rock compressive strength, m_b, s, a = material constants, m_i = material constant of intact rock, E_{rm} = deformation modulus of rockmass.

Using the above equations design rockmass parameters such as cohesion, friction angle, and rockmass modulus shall be calculated and can be used for underground structural design.

4.7 Summary

The geotechnical investigation planning and interpretation of data described in this paper can ensure that relevant geotechnical information and data are available at the various stages of the project. The ultimate aim of establishing the soil, rock, and groundwater conditions, determining the properties of the rock and gathering additional relevant knowledge about the site, can be largely achieved through such multiphase site investigations. Understanding and critically interpreting the gathered investigation data will increase the confidence level during the design and construction phases of any underground structure. A comprehensive geotechnical investigation should be able to provide the information (1) to develop stratigraphy (existing geology) based on soil/rock weathering; (2) to superimpose the geological features onto the geology such as joint sets, faults, foliation planes, etc.; (3) to appraise the groundwater condition and flow through fractures and joints; (4) to determine the existing in situ stress conditions for avoiding anticipated failures due to excavation; and (5) to estimate the quantitative rockmass parameters that are critical for design through rockmass classification systems such as Q and GSI. This entire process of site investigation and data interpretation will help in designing adequate support system for any complex underground structures with economical design and avoid sudden surprises during construction which may prove unsafe and costly.

References

ASTM D4428M Standard test methods for cross-hole seismic testing

ASTM D4879 Standard guide for geotechnical mapping of large underground openings in rock

Barton N, Lien R, Lunde J (1974) Engineering classification of rock masses for the design of tunnel support. Rock Mech 6: 189–236. https://doi.org/10.1007/BF01239496

Barton N, Grimstad E (2014) An illustrated guide to the Q-system, vol 79, 44p

Bell FG (ed) (1992) Engineering in rockmasses. Butterworth-Heinemann Ltd., Oxford, 416p

CIRIA C652 Geophysics in engineering

DIPS user manual Rocscience Inc.

Hoek E (1994) Practical rock engineering

Hoek E, Brown ET (1997) Practical estimates of rock mass strength. Int J Rock Mech Min Sci Geomech Abstr 34(8):1165–1186.

Performance of Stone Columns in Soft Clay—A Comparative Study of Bearing Capacity of Soil Estimated Using Code Method, Actual Field Load Tests, and Observations from Numerical Model—A Case Study

V. S. Parameswaran and Venkata Krishnaiah Yeddala

Abstract Assam cancer care foundation (ACCF) proposed the construction of hospitals and related facilities across 18 locations in the state of Assam, India. In this paper, study on performance of stone columns in soft clayey soils at **Jorhat** site, one among the 18 locations is discussed. At this site, ground improvement by vibro stone columns of 900 mm diameter were installed at spacing of 2.0 m C/C in square pattern up to a depth of 9 m below founding level by dry bottom feed vibro displacement method to improve the safe bearing capacity of in situ soil from 70 to 140 kPa to support hospital building, LINAC block, and service building, by method of reinforcing the ground as they provide increased bearing capacity and reduce the foundation settlements. Stone columns were designed based on guidelines given in IS 15284(Part-I)-2013. The reliability of any method can be justified based on their ability to predict safe bearing capacities. Design is validated by two initial field load tests (single and group column test) and three routine load tests (single column test) as per guidelines given in IS code. Results showed 20% higher than estimated safe bearing capacities. Numerical modeling was also performed using PLAXIS-2D program. Deformations and stability checked using software and results were compared with actual settlements estimated empirically along with the field load tests. Comparison of results shows theoretical, field load test results hold good with numerical analysis results.

Keywords Vibro stone column · Single column and group column load tests · Silty clayey · Dry bottom feed vibro displacement method

V. S. Parameswaran · V. K. Yeddala (✉)
L&T Constructions, Chennai, India
e-mail: yvkk@lntecc.com

V. S. Parameswaran
e-mail: parameswaranvs@lntecc.com

B. Adimoolam et al. (eds.), *Deep Foundations for Infrastructure Development in India*, Lecture Notes in Civil Engineering 315, https://doi.org/10.1007/978-981-19-8598-0_2

1 Introduction

M/s. L&T Constructions, Buildings and factories division is awarded a contract by Assam Cancer Care foundation (ACCF) to construct hospitals and related facilities across 18 locations in the state of Assam. The 18 locations include expansion of State Cancer Institute of Assam. The entire state of Assam lies in the Seismic Zone-V and subjected to frequent earthquakes. **Jorhat** is one among the 18 locations of Assam at which mainly the Hospital building (G+3 Stories), LINAC block (G+1), and service building and Ramp (G+3 Stories) are proposed to be constructed (Fig. 1).

Building Foundation System

Buildings	No. of floors	Foundation type	Type of structural system
Hospital block	G+3	Raft—60 m × 45 m	RCC beams and slabs with columns/walls with base isolators
LINAC block	G+1	Raft—36 m × 20 m	Same structural system, but w/o base isolators
Service building	G+3	Raft—48 m × 36 m	Same structural system, but w/o base isolators

The minimum safe bearing capacity of 14 T/m^2 is required for structure at depth of 3 m below EGL (existing ground level).

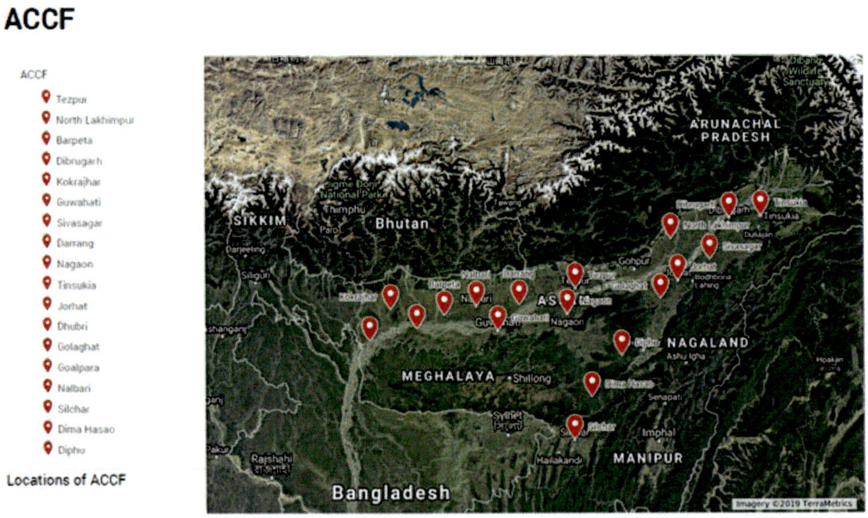

Fig. 1 Locations of Assam Cancer Care Hospitals

Table 1 Generalized soil profile

S.no	Soil strata	Depth range below NGL (m)	Field SPT-N value
1	Filled up soil	0.0–1.2	–
2	CH	1.2–5.0	5
3	CH	5.0–10.0	5
4	CH	10–13.5	6
5	SM	13.5–16.0	57
6	CH	16.0–20.0	6
7	CH	20.0–25.0	14
8	SM	25.0–30.45	31

2 Site Description and Sub-soil Profile

The proposed site is a low-lying area with high water table, as Brahmaputra River is close to site. To assess the subsoil conditions, the soil exploration is carried out at the site up to a maximum depth of 30 m. It is observed that the proposed site is underlain by top 1 m filled up soil followed by firm silty clay (N < 6) up to 13.50 m below EGL. This layer is underlain by dense silty sand (N < 57) up to 16.0 m followed by silty sand with (N > 30) up to the final exploration depth. At the time of soil investigation groundwater table observed between 0.0 and 1.2 m below the ground level. The generalized soil profile based on bore-logs for design and analysis is shown in Table 1.

3 Seismic Performance of the Ground

Subsoil being firm silty clay in nature with percentage fines more than 90% and the proposed site falls under seismic zone V (zone factor of 0.36) as per IS 1893 (Part 1) (2016) with average earthquake magnitude of 7.0, soil did not undergo liquefaction during the event of an earthquake due to higher percentage of fine content. Hence, soil is classified as non-liquefiable.

4 Safe Bearing Capacity and Settlement of In Situ Soil

Safe bearing capacity of in situ soil calculated as per IS 6403 (1981) which was 70 kPa is less than the required safe bearing capacity for raft foundation. For the raft foundation with the load intensity of 140 kPa, the estimated settlements have exceeded the allowable limit of 75 mm.

5 Objective of Ground Improvement

Soil improvement is required to achieve the bearing capacity and to restrict total and differential settlements to acceptable magnitudes under the proposed structural loading and shall provide acceptable long-term performance of treated ground.

6 Design Challenge

It is observed that the settlements exceed the allowable settlement of 75 mm for raft foundation, net safe bearing capacity is being lesser than required and hence the ground improvement is required to increase the allowable bearing capacity to 140 kPa as well as the settlements shall be within the limits.

7 Design of Ground Improvement Scheme

To address the bearing capacity and settlement issues, ground improvement in the form of vibro stone columns of 0.9 m diameter with a spacing of 2 m c/c with square pattern was proposed all over the building foot print area based on IS code. A sand blanket layer of 500 mm was provided over the installed stone columns and compacted in layers to a relative density of 80% for proper load distribution and also function as a drainage layer for pore water pressure dissipation. A typical sketch of stone column arrangement is shown in Fig. 2.

Stone Column Design Parameters

a. Angle of internal friction = 40°.
b. Unit weight of column material = 20 kN/m^3.

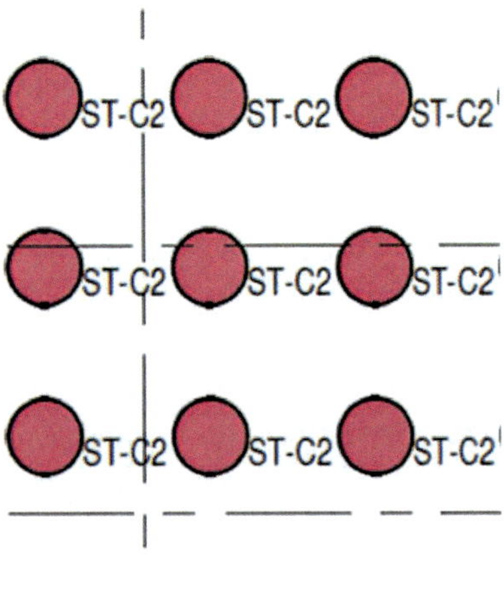

Fig. 2 Square pattern arrangement of stone columns

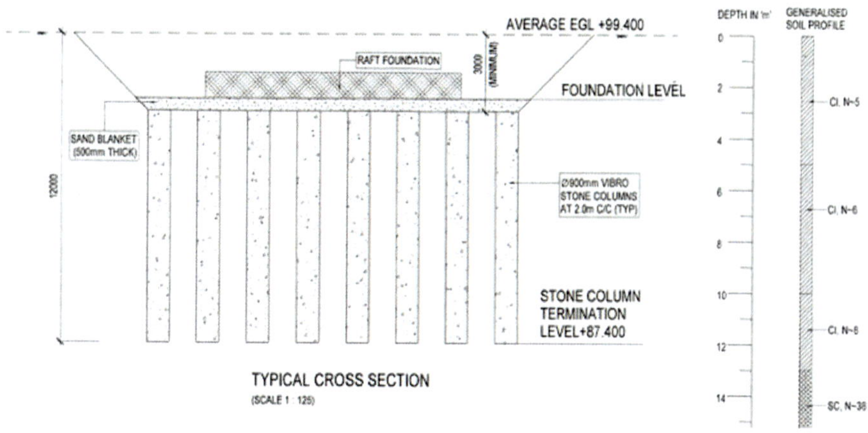

Fig. 3 Sketch showing the typical arrangement of stone column below raft

Treatment Scheme

a. Diameter of stone column = 900 mm.
b. Spacing and pattern of stone column = 2 m c/c, square.
c. Depth of foundation = 2.5 m below EGL.
d. Treatment depth = 12 m below EGL (i.e., 9 m below the founding level).
e. Sand blanket thickness = 500 mm.
f. Replacement ratio = 16% (Fig. 3).

8 Ground Improvement Methodology

At this site, dry vibro displacement bottom feed deep compaction technique was adopted to install the stone columns and to improve the load bearing capacity of the soft clayey soils. Stone columns act as a reinforcement element and increase the load bearing capacity of the soil to sustain the structural load.

While installing the stone columns by Dry vibro bottom feed method, a total of 1195 stone columns were proposed over a footprint area of 3835 m^2. One to two additional rows of stone columns were provided periphery of the raft foot print area to provide confinement for the foundation structures of this site. The muck generated during construction activities shall be minimized by proper planning and rig movement. The typical crane setup of dry bottom feed method of installation of stone columns is shown in Fig. 4.

Stone Consumption

The consumption of stone backfill shall not be less than the theoretical volume of the column (calculated based on the column diameter specified) plus additional volume due to compaction effort and losses during installation. The rate of stone consumption shall be assessed by averaging the material used in 10 adjacent compaction points.

Fig. 4 Typical crane setup by dry bottom feed method (Courtesy: Keller)

Table 2 The stone column aggregate consumption

Material reconciliation				
S. no	No. of stone column achieved	Theoretical aggregate consumption (Cum) (dia. of stone column 900 mm and length of stone column 9 m below foundation level)	Actual aggregate consumption (Cum)	Difference
1	1195	6839	7707	12.69%

As per our records, stone consumption is 15–20% more than theoretical volume. Table 2 indicates the stone column aggregate consumption.

9 Quality Assurance/Quality Control of Stone Columns

Quality Check for Stone Aggregates

As a minimum, the following controls are adopted to maintain a good quality of stones for the vibro stone column works:

- The aggregates should be chemically inert, hard, and resistant to breakage.
- Well-graded gravel with gradation ranging from 50 to 12 mm is used for vibro displacement stone columns. The gradation is chosen in such a way the voids between the stones is filled by the in situ material which is suspended in water

to form a rigid column. The gradation of stone aggregates is given in Table 3. Aggregate quality requirement for vibro stone column work is given in Table 4.

The stone columns are terminated at depth 12 m below the natural ground level depending on the penetration resistance achieved through the vibrator. The degree of compaction is measured in terms of electric current amperage. During vibro compaction, quality control can be implemented to ensure vibro compaction to achieve desired performance through amperage measurement of the vibrator. The quality control items include the location of compaction points, depth of penetration, vibratory power consumption or electric current amperage, frequency and duration of penetration, compaction, and extraction, depth of obstruction encountered if any, and ground subsidence, all these are measured through M4 graphs which are generated automatically during the installation.

Execution of vibro stone column works requires quality which is assured by implementing various measures at different stages.

These are

- During construction—monitoring and recording of process parameters.
- After construction—initial stone column test and routine stone column test.

Automated Real-Time Monitoring

The monitoring of each stone column is performed using an automated computerized recording device. This instrument yields a computer record of the installation process in a continuous graphical mode plotting depth versus time and power consumption (compaction effort) versus time. The information provided includes

- Stone column reference number.
- Date of installation.
- The period required for installation.

Table 3 Gradation for vibro dry bottom feed stone columns

Sieve size (mm)	% passing, by weight
50	100
37.5	90–100
25	30–80
20	10–50
14	0–2

Table 4 Quality requirements for stone column aggregates

Type of tests	Code	Criteria
Crushing value	IS 383	<30%
Impact value	IS 383	<30%
Stone gradation	IS 383	75–12 mm

- Maximum depth.
- Compaction effort during penetration and compaction process.

These graphs are the main quality control and monitoring tools during the installation process.

10 Validation of Ground Improvement

Numerical Modeling

One of the commonly used design approaches (IS 15284) (2003) was followed for design of stone columns in clayey soils in India. While the same design approach is adopted in this project, two-dimensional (2D) finite element analyses were also carried out to assess the settlement and bearing capacity assessment of the soil. The finite element model is constructed based on the actual design stone column layout.

In this work, two-dimensional finite element analysis of stone columns is analyzed using PLAXIS (2D) under the structural loads. The Mohr–Coulomb models are used to simulate the behavior of clayey soil and stone column, respectively. While linear elastic model is used to simulate the behavior of structures (raft foundation). By taking longitudinal section of hospital building in the finite element model the settlement contours obtained from the finite element model are corresponding to loads of 140 kPa presented in Fig. 5.

Field Load Tests:

To assess the extent of improvement, field tests such as load tests were conducted after the completion of ground improvement works to verify quality and effectiveness. The plate was designed to simulate loads imposed by sand bags which shows that under working plate loading pressure of 140 kPa, the measured plate settlement was only about 7 mm. At plate loading pressure of 210 kPa, the measured plate settlement increased to about 12.34 mm. The finite element model (FEM)-predicted plate settlement was significantly larger than that observed in the field because field load tests were conducted on single column and group columns only. This is likely due to the scenario that the actual stiffness of the stone columns is significantly higher than the adopted design stiffness. This is considered logical as in many ground improvement works, a lower bound design value is adopted and the design value typically represents the lower bound magnitude obtained in the field. This may also be attributed to the fact that soft clay properties could have improved by some extent after the installation of the stone columns. Both strength and stiffness of the soft clay are expected to have increased and these beneficial effects are not quantified in the analyses.

After installing stone columns, field load tests such as initial stone column load tests (i.e., single and group load tests) and routine stone column load tests (i.e., single column load tests) were carried out at site for assessing the extent of improvement

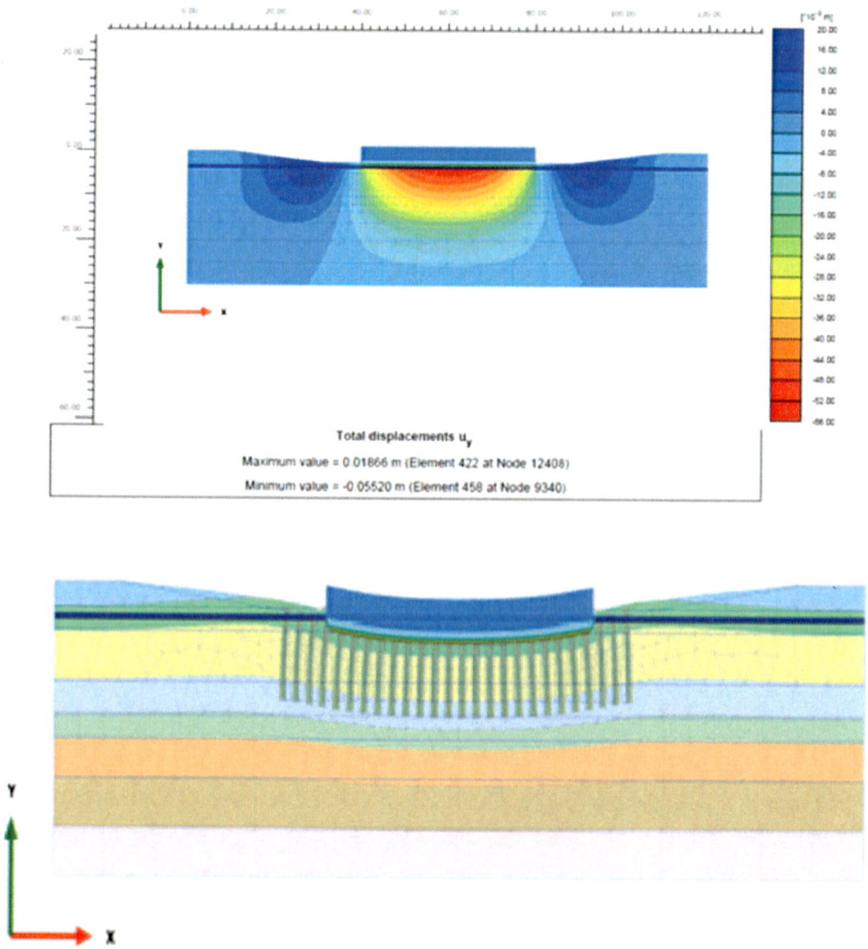

Fig. 5 Settlement contours for 140 kPa load intensity

and to validate design intent. Three routine load tests, one initial single column load test, and one group column load test were conducted.

Acceptance Criteria Initial Load Tests as per IS 15284 (2003)

1. Initial single load test: allowable settlement is 12 mm @ design load and plate size is 2.26 m.
2. Initial group load test: allowable settlement is 30 mm @ design load and plate size is 4.52 m.
3. Routine load test: allowable settlement is 12 mm @ design load and plate size is 2.26 m.

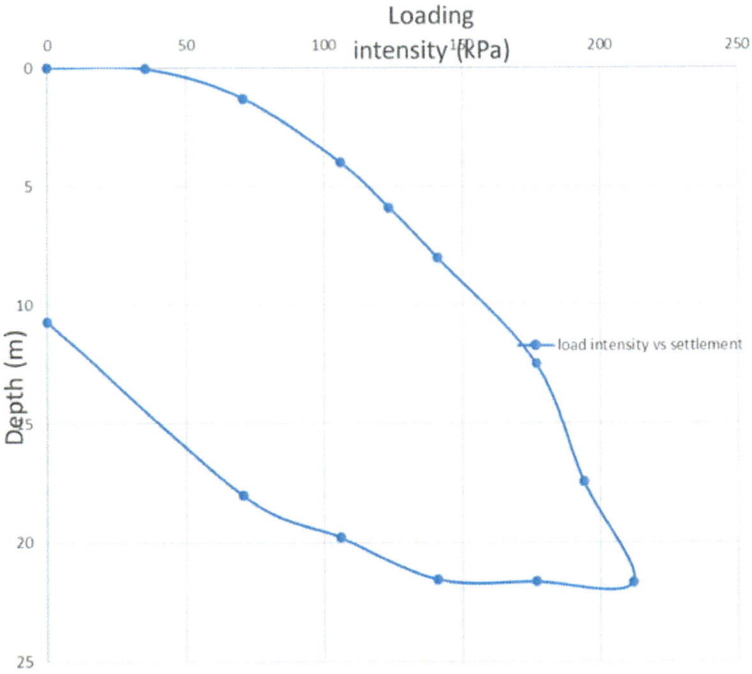

Fig. 6 Load intensity versus settlement curve of initial single load test

From Fig. 6, at design load intensity of 140 kPa, settlement is 8 mm, settlements were within permissible limits at design load as per IS 15284 (2003), hence ground improvement with stone column is safe to take structural loads.

From Fig. 7, at design load intensity of 210 kPa, settlement is 12 mm, settlements were within permissible limits at design load as per IS 15284 (2003), hence ground improvement with stone column is safe to take structural loads.

From Fig. 8, at design load intensity of 140 kPa, settlement is 7 mm, settlements were within permissible limits at design load as per IS 15284 (2003), hence ground improvement with stone column is safe to take structural loads.

11 Conclusion

In this study detailed in the paper, the designers recommend the ground improvement technique using stone columns as an effective method to increase safe bearing capacity of clayey soils with higher fine percentages and it is obvious from the load test results that the proposed scheme of stone column group demonstrated to sustain loading pressure of 140 kPa within acceptable settlement. Subsequent operation of reinforcing ground with stone column has strengthened the existing ground to meet the target performance criteria. The replacement ratio of 16% considered was

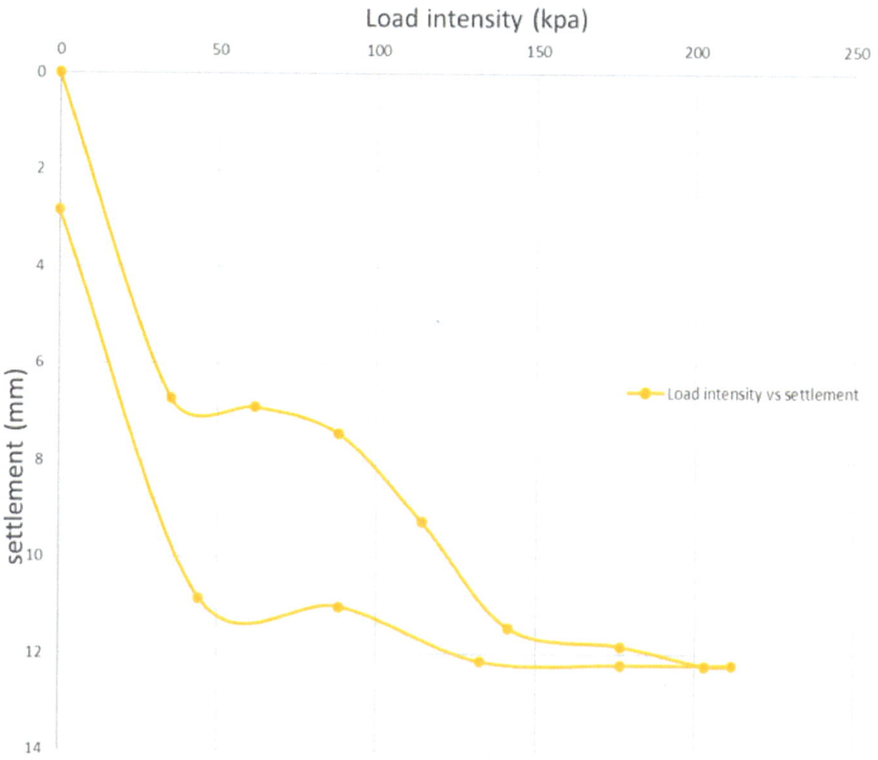

Fig. 7 Load intensity versus settlement curve of initial group load test

successful; however, as fine content varies throughout site, it is suggested to consider area replacement ratio in the range of 18–20% for significant results. Bentley Software Plaxis 2D was used to access the total deformation of raft foundation with soil structure interaction and group effect of stone columns on raft settlement behavior, since it is important for stability of structure and limitations of IS codes/theoretical methods of approach was resolved by modeling stone columns and raft foundation in Plaxis 2D software. From analysis of the Plaxis, settlement of raft is 55 mm which is less than IS code specified allowable settlement 75 mm. Hence, it is concluded that numerical analysis gave close realistic settlement which nearly matches the empirical approaches along with structure interaction and group effect of stone columns consideration on raft settlement.

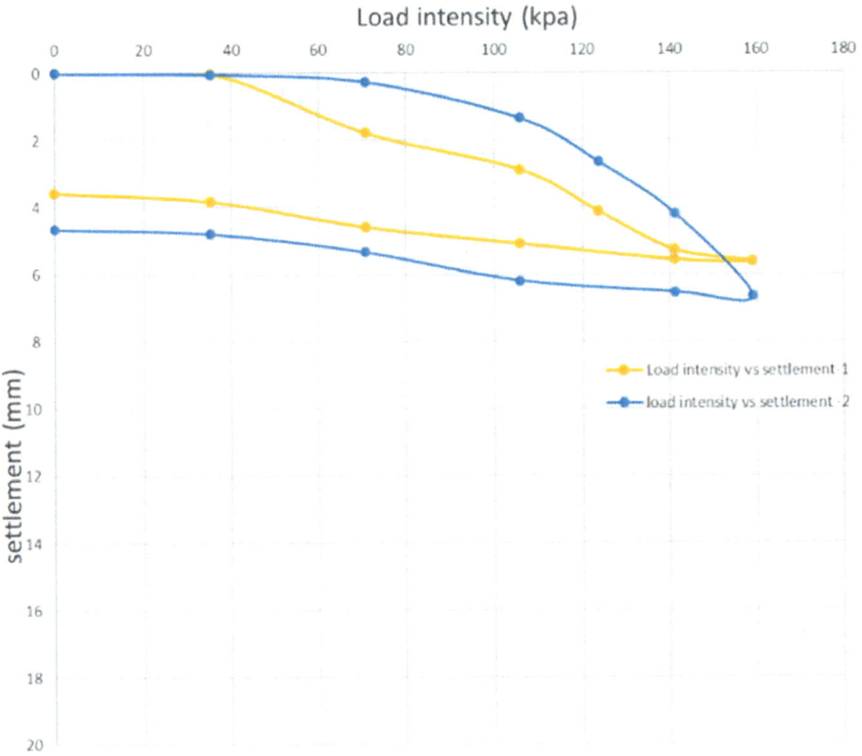

Fig. 8 Load intensity versus settlement curve of routine group load test

12 Site Photos of Stone Column Load Test

See Fig. 9.

Fig. 9 Typical routine stone column load test arrangement

Acknowledgements Authors are also thankful to **Dr. Subhadeep Banerjee,** Associate Professor of Geotechnical Engineering, Indian Institute of Technology Madras, Chennai for his support and suggestions. He wholeheartedly gave appointments for our clarifications and doubts despite his busy schedules.

The authors acknowledge the help rendered by the geotechnical team of **M/s. Keller ground engineering Pvt. Ltd** for the successful execution of stone columns and conduction of post field tests such as SPT and ECPT.

References

IS: 1893 (Part-1) (2016) Criteria for earthquake resistant design of structures – part 1: general provisions and buildings. Bureau of Indian Standards, New Delhi

IS: 6403 (1981) Code of practice for determination of bearing capacity of shallow foundations

IS 15284: Part 1 (2003) Design and construction for ground improvement – guidelines – stone columns

Use of Reliability-Based Approach to Determine Geotechnical Parameters of Soil Site

Gurpreet S. Bhatia and G. Prabhakar

Abstract Proper geotechnical investigation plays an important role in suitably addressing the geotechnical challenges faced during construction stages. However, improper investigations lead to shocking results both financially and technically. Some of the most important geotechnical investigation tests like the Standard Penetration Test (SPT), Pressure Meter Test (PMT), etc. which are very much important in determining the foundation parameters, require too much manual intervention. These factors govern the workmanship of geotechnical investigation and the results we obtain from them. This leads to uncertainty in the results of these investigations. Generally, a number of tests are conducted below a particular structure and their average results are used for the design of foundations or other underground utilities. Against the backdrop of the fact that there are no stringent guidelines on the extent of geotechnical investigations to be carried out below critical structures like defense facilities, nuclear facilities, etc., this method of averaging may lead to obnoxious results. The greatest drawback of this approach is that we are not using sufficient data to characterize the soil profile of any particular area. In case some more investigations are done in that area, the results may change drastically. In this paper reliability-based approach is used to estimate the parameters obtained from SPT and PMT. The probabilistic approach is used to estimate the 95 percentile values of parameters that become input for the design of underground structures or utilities. As uncertainty exists in the evaluation of these parameters due to improper investigations, it is appropriate to evaluate these parameters based on the probabilistic approach using the best fit probabilistic distribution curve. This approach helps in the conservative estimation of geotechnical parameters below any structure with minimal failure probability.

Keywords Standard penetration test · Pressure meter test · 95 percentile values

G. S. Bhatia (✉) · G. Prabhakar
Nuclear Power Corporation of India Limited, Hisar, Haryana, India
e-mail: bhatiasgurpreet09@gmail.com

G. Prabhakar
e-mail: gprabhakar@npcil.co.in

B. Adimoolam et al. (eds.), *Deep Foundations for Infrastructure Development in India*, Lecture Notes in Civil Engineering 315, https://doi.org/10.1007/978-981-19-8598-0_3

1 Introduction

The biggest element of technical and financial risk in any Infrastructure project lies majorly in sub-strata. An inadequate or inappropriate geotechnical investigation often leads to structural failures. Due to the absence of stringent guidelines on the extent of geotechnical investigations to be carried out below critical facilities, it often leads to inadequacy of investigations that are required to appropriately map the strata available below these structures. This inadequacy in geotechnical investigations leads to a lack of knowledge about the sub-strata which further leads to faulty designs. AERB Safety Guide No (2008) illustrates the Geotechnical Aspects and Safety of Foundation for Buildings and Structures important to the safety of Nuclear Power Plants. The cost incurred at geotechnical investigations is very low, somewhere around 0.01–0.2% only when compared to the overall Project cost (Jaksa et al. 2003). This insufficient coverage often leads to uncertainties, which results in unforeseen costs and delays during the construction stage.

A reliability-based approach is required to estimate geotechnical parameters due to the large uncertainties associated with the estimation of these parameters. It is highly difficult to predict the behavior of soil due to spatial variations and the presence of local irregularities in strata even in the relatively smaller area (Lacasse and Nadim 1998). This aspect is more predominant in alluvial soils as the formation of different layers has taken place with time and every layer may behave differently leading to more uncertainties in understanding the actual soil behavior. The reliability approach provides the geotechnical engineers an edge to deal with the inherent risks associated with the investigation data and helps them to reduce the failure probability during design. This approach takes into account every data obtained and provides a value that has a specific probability of occurring based on the entire collected data under consideration. In case some very low or high value is encountered in a particular zone, then this technique will take that value also into account in the estimation of a specific percentile value unless that value is ignored, considering it an outlier.

This paper aims to establish the procedure for determining the geotechnical parameters from the conventional investigation methods using a reliability approach. These parameters can then be used for design purposes with higher confidence level as some amount of uncertainties associated with them will get eliminated. The focus will be to derive parameters from the two most important field tests in geotechnical engineering, i.e., SPT and PMT. Parameters obtained from these tests can then be utilized to design the structures with a reduced level of failure probability.

2 Geotechnical Investigation to Determine Field Parameters

One of the most important field tests that is widely used worldwide to determine soil properties is the Standard Penetration Test (SPT) (IS 2131 1981). SPT provides data that can be used as input to determine various soil properties like liquefaction potential, bearing capacity (IS 6403 1981), relative density, angle of shearing resistance, etc. Conduction of SPT at the site requires skilled manpower. There is too much manual intervention in the method generally being adopted to determine SPT values at the site. Due to the absence of skilled workmanship and too much manual intervention, there are possibilities that the data obtained may have a large number of uncertainties and may not represent the actual soil behavior. In order to minimize the human/machine error as far as possible, all tests were performed by the same operator/technician using the same equipment. All equipment and accessories were calibrated at regular intervals to avoid any error arising out of the use of faulty equipment. All Standard penetration tests were performed using safety type SPT hammer and no in-situ hammer efficiency was measured for the type of hammer used. However, it may be noted, that by using automatic trip SPT hammers and by performing hammer efficiency at the site, many uncertainties and errors associated with measuring site data can be minimized further.

Similarly, the pressure meter test is also one of the important tests that provide information on the Modulus of Elasticity of soil which is very much essential in designing sub-structures and estimating settlements of soil. As compared to SPT, the pressure meter test is even more complex to conduct and not everyone has the required skills to conduct this test. Generally, the Menard-type pressure meter is used for the estimation of the modulus of elasticity of soils as per the guidelines given in ASTM-D 4719-20 (ASTM D4719 - 20 en Standard Test Methods for Prebored Pressuremeter Testing in Soils 1.1) or ISO 22476-4 (2012). One of the most critical parts in the conduction of a pressure meter test is the preparation of the test pocket. With the presence of a water table and loose soil this process becomes even more complex. In case there are some issues with pocket preparation, even the most skilled person will not be able to conduct the test in a proper manner. This will eventually add uncertainties in the field results and may not provide a true resemblance of actual soil properties. As far as possible, the results of the pressure meter test which seems to represent ambiguous data due to reasons attributable to improper pocket preparation, test pocket failure, loosening of soil while drilling, etc. have been ignored and re-test was performed at slightly lower depth to re-assess data of specific depth for that particular borehole.

The probabilistic analysis will be carried out on the available field data from these tests in order to determine the 95 percentile values of these parameters. Higher percentile values indicate a higher confidence level with which this parameter has been obtained from the available raw data (Haldar and Mahadevan 2000). Data from geotechnical investigations carried out in approximately 300 m $\times$ 300 m area in the Northern region of India has been used to conduct these studies. A total of 20 boreholes were drilled to conduct both SPT and PMT. These two tests were performed in the same borehole at different depths. SPT was conducted at an interval of 3 m along the depth of the borehole while PMT was conducted at an interval of approximately 5 m. Results obtained from these tests show that there is large variability in the results. Generally, the averaging technique is used to design foundations based on SPT data. SPT results from different boreholes at particular depths are averaged and results are then used for further design calculations. In case some values are greater than 50% of the average value, then those results are neglected in calculating the average. This same methodology has been adopted to estimate 95 percentile values from the test results of SPT. In the case of PMT, all field values have been considered and no value has been neglected or capped for estimating 95 percentile values. However, there is very high variability in the results of PMT mainly due to the fact that it captures the soil property of small pocket very precisely. There are certain uncertainties associated with the results of PMT mainly due to the testing method and condition of the test pocket in which the test was conducted.

Figure 1 shows the depth-wise distribution of PMT values as obtained from a field test using the Menard Pressure meter apparatus. Figure 2 shows depth-wise distribution of SPT values obtained in different boreholes. From both figures, it can be observed that there is too much variability in the obtained field data.

The disadvantage of the averaging technique is that we are not taking into consideration the variability of data appropriately. This method may not provide the most conservative results which might be suitable for designing any critical facilities. Similarly, if minimum value is considered to design critical facilities in a conservative way, it might also lead to erroneous results. Taking a minimum value for calculation is not recommended because this minimum value is coming from the data available based on limited geotechnical studies. In case some more investigations are carried out in a similar area, there is a fairly high probability that this minimum value will

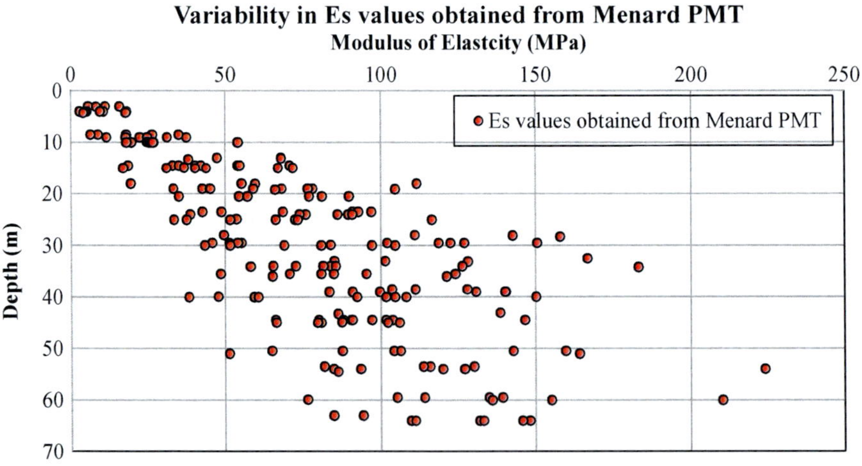

Fig. 1 Variability in values of Static Modulus of Elasticity along the depth of boreholes

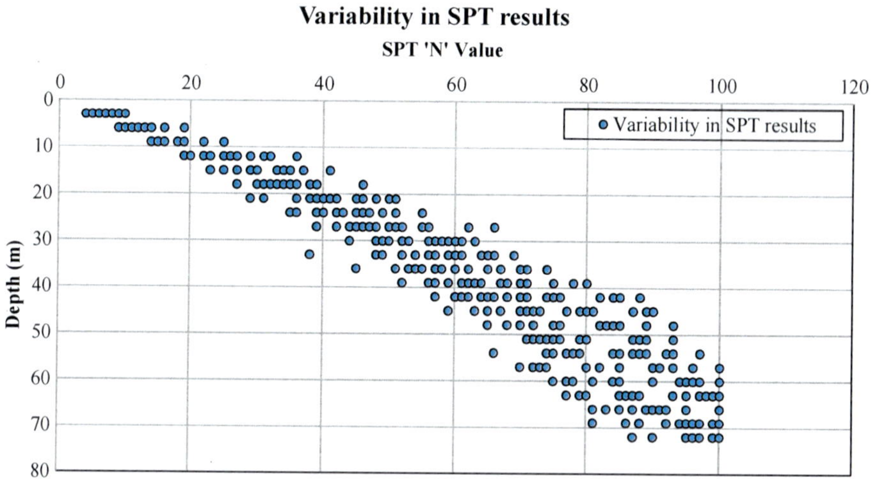

Fig. 2 Variability in SPT values along the depth of boreholes

get changed, and then we have to again re-design everything considering the new value. In order to avert such situations, the probabilistic approach is considered to be more effective in the determination of design parameters from field data. Obtaining 95 percentile values of these parameters will give substantial confidence that even if some more investigations are carried out there is only a 5% probability that we will get values lower than already obtained values using the probabilistic approach.

3 Assessment of 95 Percentile Values

The approach will be to identify the best fit function which represents this data at each depth. Data from available 20 boreholes are analyzed at each depth and different probabilistic distribution curves (like normal, lognormal, and gamma) are used to find the best suitable curve for each parameter at each depth under consideration. In order to find out the 95 percentile value of data from these distributions, it is first required to plot the PDF and CDF of the random data using all the probability distribution functions. In case random data satisfies all the probability functions, it is then required to choose the best function among the available functions that best suits the randomness of data. Comparing probability functions with each other can be done by performing statistical tests like the Chi-square test, Kolmogorov–Smirnov test, or Anderson–Darling test.

In this paper, the K–S test has been used to identify the best probabilistic function out of the available functions for raw data at each depth (Haldar and Mahadevan 2000). The K–S test can be applied on the small number of data effectively as compared to the Chi-square test. Moreover, the advantage of the K–S test is that it is not necessary to divide the data into intervals; thus the error or judgment associated with the number or size of the interval can be avoided.

The K–S test compares the observed cumulative frequency and the CDF of an assumed theoretical distribution. The first step is to arrange the data at a particular depth in increasing/ascending order of their absolute value. Then the maximum difference between the two cumulative distribution functions of the ordered data can be estimated as

$$Dn = \max|Fx(x_i) - Sn(x_i)| \tag{1}$$

where $Fx(x)$ is the theoretical CDF of the assumed distribution at the ith observation of the ordered samples x_i, and $Sn(x_i)$ is the corresponding stepwise CDF of the observed ordered samples. $Sn(x_i)$ can be estimated as

$$Sn = \left\{ \begin{matrix} 0, x < x_i \\ \frac{m}{n}, x_m \leq x \leq x_{m+1} \\ 1, x \geq x_n \end{matrix} \right\} \tag{2}$$

Mathematically, Dn is a random variable and its distribution depends on the sample size n. The CDF of Dn can be related to the significance level α as

$$P\left(Dn \leq D_n^{\alpha}\right) = 1 - \alpha \tag{3}$$

And the D_n^{α} values at various significance levels can be obtained from a standard mathematical table. Thus, according to K–S test, if the maximum difference Dn is less than or equal to the tabulated value of D_n^{α}, the assumed distribution is acceptable at the significance level α.

The probability density (PDF) of the normal distribution is expressed as

$$fX(x) = \frac{1}{\sigma_x\sqrt{2\pi}}\exp\left|-\frac{1}{2}\left(\frac{x - \mu_x}{\sigma_x}\right)^2\right|, -\infty < x < +\infty \tag{4}$$

The corresponding CDF can be expressed as

$$FX(x) = \int_{-\infty}^{x} \frac{1}{\sigma_x\sqrt{2\pi}}\exp\left|-\frac{1}{2}\left(\frac{x - \mu_x}{\sigma_x}\right)^2\right|\mathrm{d}x \tag{5}$$

where μ_x is the mean or expectation of the distribution; σ_x is the standard deviation; σ^2 is variance.

Probability Density function (PDF) of the Lognormal Distribution can be calculated as

$$fX(x) = \frac{1}{\sqrt{2\pi}\zeta_x x}\exp\left[-\frac{1}{2}\left(\frac{\ln x - \lambda_x}{\zeta_x}\right)^2\right], 0 \le x \le \infty \tag{6}$$

where λ_x and ςx are two parameters of the lognormal distribution. The lognormal variable has values ranging from 0 to ∞. Two parameters of the lognormal distribution can be calculated from the information on the two parameters of the normal distribution, the mean (μ_x) and the standard deviation (σ_x) of the sample population. It can be estimated as

$$\lambda x = E(\ln x) = \ln \mu x - \frac{1}{2}\zeta_x^{\ 2} \tag{7}$$

$$\zeta_x^{\ 2} = \mathrm{Var}(\ln X) = \ln\left|1 + \left(\frac{\sigma_x}{\mu_x}\right)^2\right| = \ln(1 + \zeta_x^{\ 2}) \tag{8}$$

PDF of the Gamma distribution can be calculated as

$$f(x) = \frac{\left(\frac{x-\mu}{\beta}\right)^{\gamma-1}\exp - \left(\frac{x-\mu}{\beta}\right)}{\beta\Gamma(\gamma)} x \ge \mu;\ \gamma, \beta > 0 \tag{9}$$

where γ is the shape parameter, μ is the location parameter; β is the scale parameter, and Γ is the gamma function having the formula

$$\Gamma(a) = \int_0^{\infty} t^{a-1}e^{-t}\mathrm{d}t \tag{10}$$

CDF of the gamma function can be calculated as

$$F(x) = \frac{\Gamma_x(\gamma)}{\Gamma(\gamma)} x \ge 0;\ \gamma > 0 \tag{11}$$

where Γ is the Gamma function defined above and $\Gamma x(a)$ is the incomplete gamma function. The incomplete gamma function has the formula

$$\Gamma x(a) = \int_0^{x} t^{a-1}e^{-t}\mathrm{d}t \tag{12}$$

There are several computational methods for the K–S test. In this study, K–S test static version is used. First, sort the data. Then establish the assumed distribution and estimate its parameters. Then obtain both the theoretical (assumed CDF) distribution (FX) as well as empirical (Sn) at each data point. Since K–S is a distance test, it is required to find the maximum distance $|Fx - Sn|$ between the theoretical and empirical distribution. Its two basic functions are described as

$$FX(em) = PX(X \leq em) = \text{CDF}(em) \tag{13}$$

FX(*em*) is the assumed cumulative distribution function evaluated at *em* and *Sn*(*em*) is the empirical distribution function obtained by the proportion of the data smaller than *em* in the data set of size *n*.

$$Sn(em) = i/n;\ i = 1, 2, \ldots n. \tag{14}$$

Then, define $D+ = Sn - FX$ and $D- = FX -$ Sn-1 for every data point *em*. The K–S static is

$$D = \text{maximum of all} D + \text{ and } D - \ (\geq 0);\ \text{for } em = 1, 2 \ldots n. \tag{15}$$

The K–S logic is, if the maximum departure between the assumed CDF and empirical distributions is small, then the assumed CDF will likely to be correct. But if the discrepancy is "large" then the assumed *FX* is likely not the underlying data distribution.

For a 95 percentile level and 20 sample points, $Dn\alpha$ is calculated from the standard table for the K–S test.

This assessment was done for each depth of the data set of SPT and PMT individually. From the assessment, 95 percentile values are obtained for SPT and PMT at each depth. As a comparative study, apart from 95 percentile values of parameters, 50 percentile and 98 percentile values were also obtained and plotted on the same curve.

95 percentile values of these parameters have been obtained at each depth at which the investigation was carried out and the same has been compared with average values in Figs. 3 and 4 for SPT and PMT, respectively.

From Figs. 3 to 4 it can be observed that higher percentile values give more conservative results as failure probability decreases with an increase in percentile values.

As can be seen from Fig. 4, there are too many variations in the values of Static Modulus of Elasticity as compared to the results of SPT obtained from different percentiles. This is mainly due to the fact that results of PMT obtained from field tests already had inherent variations with too much difference between the minimum and maximum values at similar depths in different boreholes. This variation resulted in lower values of PMT as obtained from different percentiles as compared to the average value. This shows the impact of the reliability technique in the estimation of certain parameter which contains too much variability.

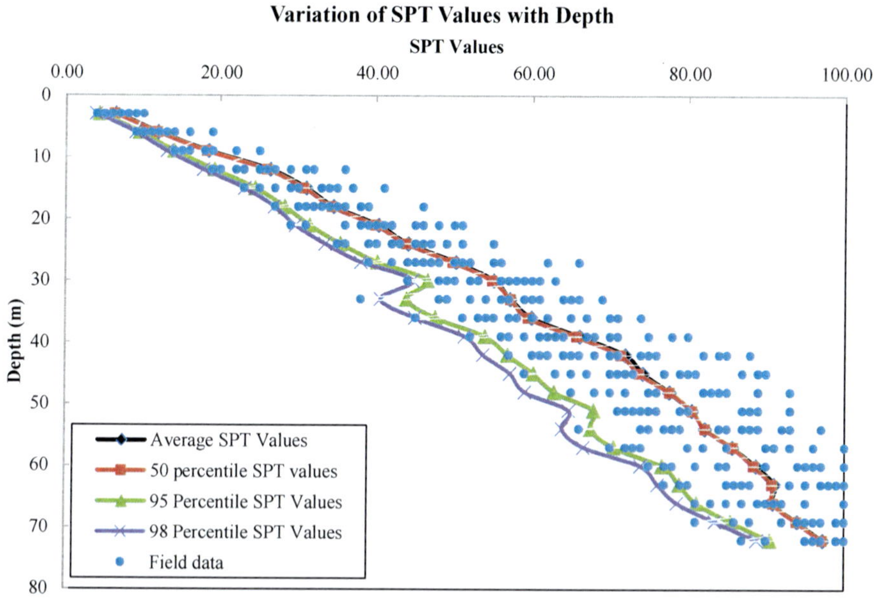

Fig. 3 Variation in SPT values along the depth obtained using different percentile values

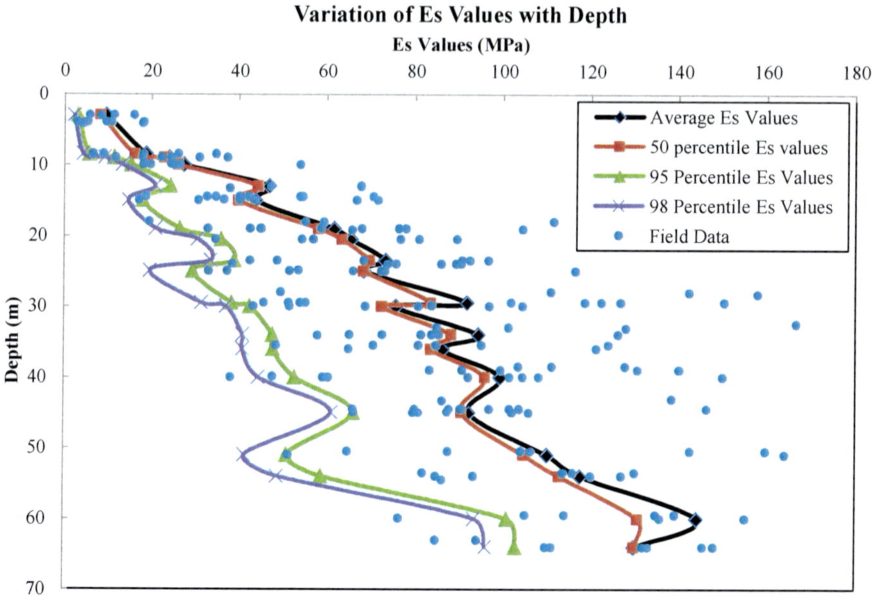

Fig. 4 Variation in Es values along the depth obtained using different percentile values

4 Conclusion

Standard penetration test (SPT) and Pressure meter test (PMT) were conducted in 20 boreholes up to a maximum depth of around 80 m from the existing ground level. As the conduction of these tests requires too much manual intervention, some uncertainties are always associated with the results of these tests. The conventional method is to average the results of field values at a particular depth from different boreholes and use the same for further design purposes. However, the biggest drawback associated with this technique is that it does not take into account the data variability appropriately. There are chances that in case some more investigations are carried out than these values may change drastically. To minimize human/machine error as far as possible, all tests were performed by the same operator/technician using the same equipment.

The reliability-based approach was used as a final check to determine 95 percentile values of parameters at different depths using the best probabilistic function based on available data. K–S test was used to determine the best probabilistic function out of three functions, i.e., normal, lognormal, and gamma. From the results, it can be observed that 95 percentile values give conservative results as compared to the average values of parameters. 50 percentile values give results almost similar to average values. Using 95 percentile values for design purpose helps in reducing the failure probability drastically and also reduces the uncertainties associated with obtaining these values during the investigation. Values obtained using different percentiles provide many conservative results as compared to average values especially for the data which contain too much variability. This helps in dealing with the uncertainties that may arise in case some additional investigations are carried out in the vicinity of area where investigations have already been carried out earlier.

References

AERB Safety Guide No. AERB/NPP/SG/CSE-2 (2008) Geotechnical aspects and safety of foundation for buildings and structures important to safety of nuclear power plants, February 2008

ASTM D4719 - 20, Standard test methods for prebored pressuremeter testing in soils 1.1

IS 2131 (1981) Method for standard penetration test for soils

IS 6403:1981 (Reaffirmed 2002) Code of practice for determination of bearing capacity of shallow foundations

Haldar A, Mahadevan S (2000) Probability reliability and statistical methods in engineering design. Wiley, New York

Jaksa MB, Kaggwa WS, Fenton GA, Poulos HG (2003) A framework for quantifying the reliability of geotechnical investigations. In: Kiureghian, Madanat, Pestana (eds) Applications of statistics and probability in civil engineering

Lacasse S, Nadim F (1998) Risk and reliability in geotechnical engineering. In: Proceedings of fourth international conference on case histories in geotechnical engineering, St. Louis, Missouri

Liner Piles Used as Support to Kentledge for Initial Compression Load Test

Thomas John, B. Venugopal, A. Vetriselvan, and M. Kumaran

Abstract The initial pile load tests are conducted to ensure the geotechnical capacity of the pile in a particular site condition. Normally kentledge method is adopted for the compression pile load tests. In this method the entire kentledge load is loaded above secondary beams which are supported by concrete blocks above the Natural Ground Level. So, it is important to ensure the SBC (for shear and settlement) of the natural ground. In one of our projects initially it was decided to go with the kentledge system for the initial compression load test on a 56 m long 1200 mm diameter BCIS pile. The test load and total kentledge load were 2160t and 2700t, respectively. The soil strata mainly consisted of clay with a low SPT N value and a water table almost at ground level. It was suggested to replace top soil using a Granular Sub-base (GSB) with good compaction in order to avoid excessive settlement of concrete blocks placed to support the kentledge weight. When the kentledge loading started, even before reaching 10% of the total kentledge load, an excessive settlement on the GSB layer was observed. This paper describes the problems faced during the normal kentledge method and how it was overcome using the liner pile system for initial compression load tests in this particular site condition for a higher test load.

Keywords Geotechnical capacity · Kentledge · Liner piles · Load test · Settlement

T. John (✉) · B. Venugopal · A. Vetriselvan · M. Kumaran
L&T GeoStructure, Chennai, India
e-mail: thomasjohn@lntecc.com

B. Venugopal
e-mail: bvg@lntecc.com

A. Vetriselvan
e-mail: avsn@lntecc.com

M. Kumaran
e-mail: mkn@lntecc.com

B. Adimoolam et al. (eds.), *Deep Foundations for Infrastructure Development in India*, Lecture Notes in Civil Engineering 315, https://doi.org/10.1007/978-981-19-8598-0_4

1 Introduction

A variety of methods are there to determine the load bearing capacity of piles that might be either analytical or empirical. The former requires an evaluation of soil and pile interaction along with several underlying assumptions. On other hand, the latter was based on outcomes from various in-situ test procedures. Generally pile load tests are conducted to confirm the design loads of pile obtained from empirical formulas. Before constructing the bearing piles at site, the design loads calculated theoretically need to be confirmed by conducting initial pile load test on initial load test piles. From the initial pile load test results, it was decided whether to redesign the pile or proceed construction with existing design data. Compression pile load tests normally carried out by two methods such as kentledge method or reaction pile method. This paper present the case study of initial pile load test conducted using kentledge type method where liner piles are used for kentledge support due to low SBC value of soil at existing ground level.

2 Project Scope and Stratigraphy

The project consisted of construction of pile foundation for 2 Nos of 89 m dia LNG tanks. Details of pile foundation for each tank are given in Table 1.

The soil profile of the project location shows that top 20 m is very soft clay. The site location is a filled-up area using dredged soil from sea. Due to the presence of very soft clay & loose sand strata for a depth of approximately 20 m, permanent steel casing is provided for working piles up to 24–30 m and pile is terminated in sand strata at 56 m in some locations and in other location pile is terminated at deeper depth of 61 m as dense sand strata is available after 60 m. Typical soil profile is shown in Fig. 1. Due to the presence of soft soil, it was decided to replace top 3 m soil with compacted engineering fill (GSB) for enhancing the lateral capacity of the pile.

Table 1 Details of pile foundation

Tank No	Type of pile	Pile dia (mm)	Pile depth (m)	Number of piles
Tank-01	BCIS	1200	58	242
		1200	63	194
Tank-02		1200	58	204
		1200	63	232

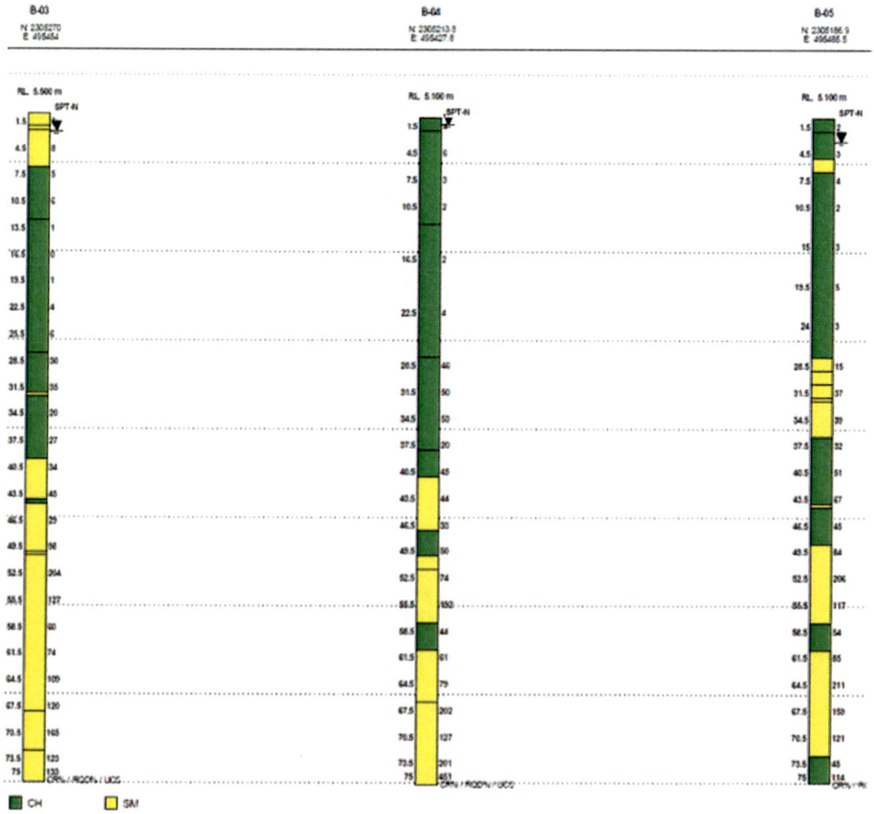

Fig. 1 Typical soil profile of site

3 Design Load and Initial Pile Load Test

Design load for the piles is given in Table 2.

To confirm the theoretical design load, it was proposed to carry out an initial compression load test at site near to the tank location. A total of 4nos of initial compression load test were proposed by the client. Due to huge test load (2160t) and availability of sufficient rebars and steel plates, kentledge system was adopted for initial compression load test as per IS 2911 Part 4: 2013. Figure 2 shows the initially proposed test scheme for the load test. The kentledge weight of 2700t is placed over

Table 2 Design load for pile foundation

Pile type	Pile dia (mm)	Pile depth (m)	Deign load (MT)	
			Safe working load (SWL)	Negative skin friction (NSF)
BCIS	1200	56/61	800	80

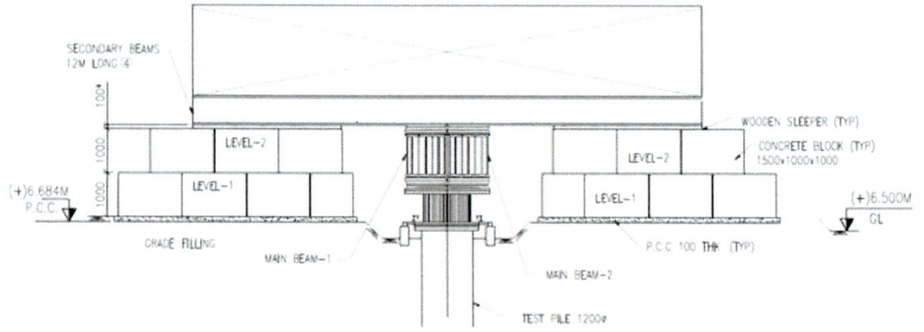

Fig. 2 Typical initial load test arrangement

48Nos of secondary beams which are supported by concrete blocks. The concrete blocks are placed over the GSB fill.

4 Site Issue in Initial Load Test and Solution

4.1 Site Issue

The base pressure acting on ground surface below concrete blocks on one side was 20 t/m^2. From theoretical calculations the settlement for 20 t/m^2 was 75–80 mm. So, a gap of 100 mm was kept between the top of the hydraulic jack and the main beam. Since the soil below the GSB fill was soft, as the kentledge loading above secondary beams, the soil below concrete blocks started to settle. Before reaching 20% of kentledge, i.e., 500 t the concrete blocks settled about 95 mm, due to which the main beam also settled, and the main beam almost touched the hydraulic jack. So, from a safety point of view as the excessive settlement of soil will lead to serious accident hence the loading of kentledge was stopped and decided to unload the materials. Figure 3 shows actual site photo during the time of kentledge loading.

4.2 Solution

To overcome this excessive settlement, it was decided to support the kentledge above the liner piles instead of using concrete blocks. The 8 mm thick 1200 mm diameter liner piles were driven to a depth of 24 m using ICE-815 vibratory hammer of capacity 72 t. 16Nos of liner piles (8Nos on either side of test pile) were driven. The nearest liner pile is at a clear distance of 2.7 m from test pile. The plan of liner pile location and connecting beams are shown in Fig. 4. The liner piles were connected at top

Fig. 3 Load test arrangement in site using concrete blocks

using 4.5 m long built-up beam consisting of 6Nos of NPB 600 beams (Represented as Beam-3 in Fig. 5). Above these built-up Beam-3, another 16 m long built-up beam consisting of 4Nos of NPB 600 beams (Represented as Beam 4 in Fig. 5). Over this Beam 4, 48Nos of secondary beams are placed. The initial load test scheme using liner pile arrangement is shown in Fig. 5.

So, with this arrangement system the kentledge load from secondary beams is transferred to a dense stratum which is at about 20 m below EGL. The top levels of four corner liners were monitored using dumpy level at regular interval till the kentledge loading of 2616 t was completed. Figure 6 shows the actual site photo for the load test arrangement as discussed. The settlement of liner piles is monitored regularly.

5 Discussion on Liner Pile Settlement During Kentledge Loading

As discussed earlier, the kentledge weight is supported by 16Nos of liner piles. After final kentledge loading load coming on each liner is 167.5 t. With such a huge load on each liner the final settlement observed in liner piles was 16 mm on left side

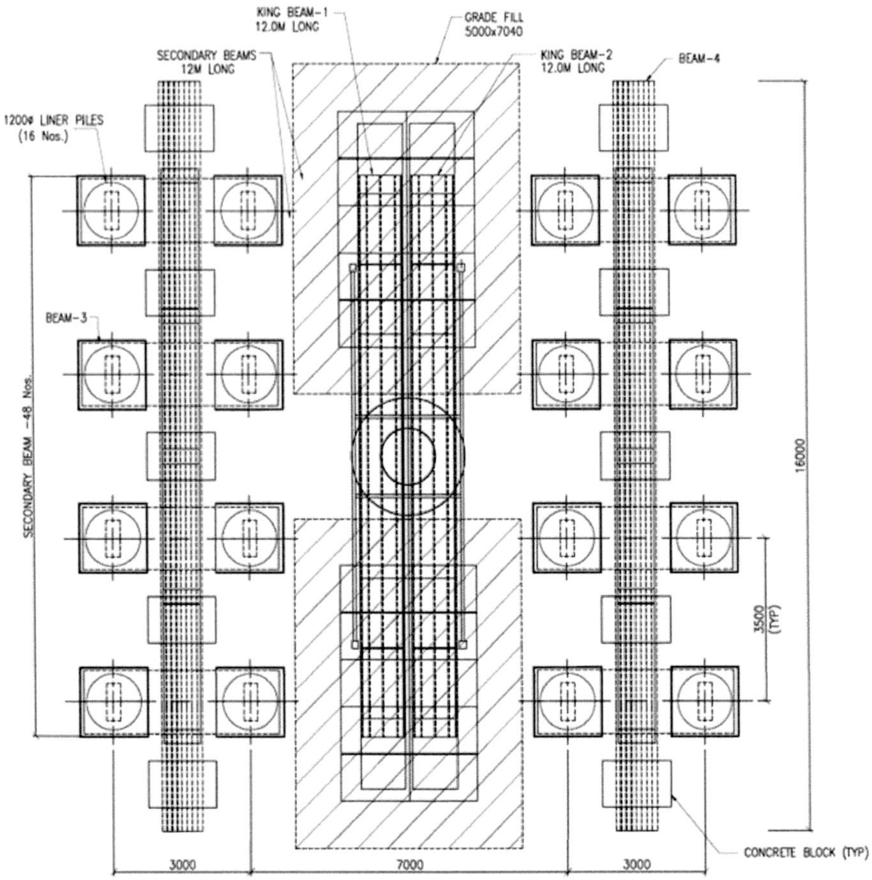

Fig. 4 Plan of load test arrangement using liner piles

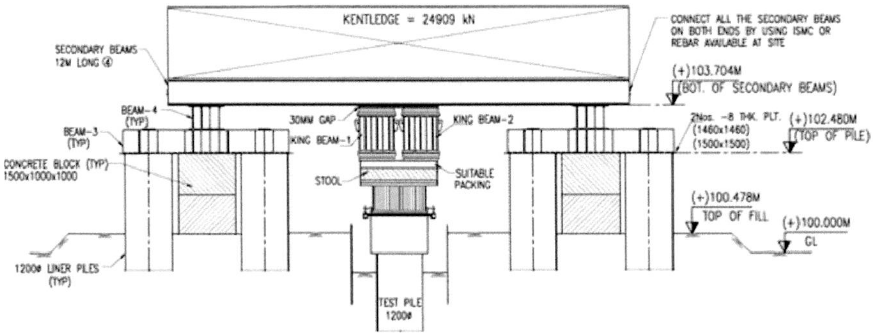

Fig. 5 Elevation of load test arrangement using liner piles

Fig. 6 Load test arrangement using liner pile as support to kentledge

and 20 mm on right side liners. Table 3 shows the left side and right side liner pile settlement at different stages of loading.

So, comparing with old load test proposal using concrete blocks as kentledge support, the liner pile system as support for kentledge was successful for heavy kentledge loads.

Table 3 Liner pile settlement at different stages of loading

Load-Settlement details for liner piles			
Total load (MT)	Load on each liner (MT)	Settlement (mm)	
		Left side	Right side
0	0.0	0.0	0.0
125.8	7.9	0.0	0.0
451.642	28.2	2.5	4.0
564.67	35.3	2.5	5.0
915.67	57.2	3.5	7.5
1283.7	80.2	4.5	8.0
1980.7	123.8	7.5	11.0
2679.7	167.5	16.0	20.5

Table 4 Test pile settlement during load test

% of load applied	Load applied (kN)	Settlement (mm)
0%	0	0
100%(SWL+2XNSF)	9720	16.35
0%	0	6.63
150%SWL+100%(2XNSF)	13,885	52.03
0%	0	36.7
200%SWL+100%(2XNSF)	17,356	123.35
0%	0	103.9

6 Discussion of Initial Pile Load Test Results

As mentioned above the test load for the initial load test is 2600t (250% SWL+100% 2xNSF). Load test was carried out in 3 cycles. First cycle with maximum load of 100% of design load applied in stages of 20% increment and each increment is maintained till rate of movement of pile top is not more than 0.2 mm/h or until 2 h have elapsed, whichever is earlier subject to minimum of 1 h (as per IS 2911 Part 4:2013). Second cycle applying load up to 150% in the same manner and in the third cycle applying load up to 250% and maintaining the final test load for 24 h.

In third cycle after applying a load of 17356kN (200% of SWL+100% of 2XNSF) the pile top settlement reached 123.35 mm (more than 10% of pile diameter) after 30 min of loading. So as per IS 2911 (Part 4): 2013, the loading was stopped and unloaded to zero.

The pile load test result was shared with the client and based on the result they have increased the number of piles in each tank as the actual capacity from the load test is less than the theoretical value (Table 4).

7 Conclusion

From the above data obtained during the load test following points can be concluded,

1. For pile load test with heavy kentledge loads, liner pile as support for kentledge weight will be safer compared to kentledge supporting on concrete blocks.
2. For sites where existing ground surface containing soft soil and having low SBC value, liner pile system is recommendable for kentledge type pile load test compared to use of concrete blocks as support to kentledge weight.

Reference

IS 2911 (Part-4) (2013) Design and construction of pile foundation – code of practice part 4 load test on piles

Ground Improvement Techniques

Development of Design Charts for Sand Compaction Pile Method of Improvement for Loose to Medium Dense Sands

N. Aarthi and G. R. Dodagoudar

Abstract The improvement of soft clay deposits using stone columns is a well-established ground improvement technique in the western world. This technique has been well documented and has proper design codes for precise execution in the field. The sand compaction pile (SCP) method is a contemporary technique for stone columns and has limited literature related to the strength characteristics of the loose to medium dense sand deposits treated with SCP. The available studies on the SCPs installed in cohesionless deposits focused on the improvement by indirectly assessing the SPT-N values pre- and post-installation of the SCPs. The widespread implementation of the SCP technique in recent years has increased the need for a more direct evaluation of the improvement. Earlier studies in this regard have revealed that the available design solutions are based on the type of installation equipment, their working efficiency, and accumulated field data. However, it is found that there are no generic design codes for the direct estimation of the ultimate bearing capacity of the SCP improved cohesionless deposits. To meet the design requirement for the SCP treated ground, a series of experimental and numerical investigations are performed in the present study to arrive at a direct framework in the form of design charts based on the pressure-settlement response of the treated ground. The developed design charts give the ultimate bearing capacity (UBC) of the treated sand deposit for the known initial relative density (RD) of the deposit, spacing and diameter of the SCPs, size of the footing, and for the specified target unit weight required for the intended application. It is concluded that the design charts will be of preliminary use to the design engineers to directly evaluate the UBC of the SCP improved loose to medium dense cohesionless deposits as part of the SCP method of ground improvement. It is expected to have more field-scale experiments and in-depth analysis before implementing these charts for actual field execution.

Keywords Sand compaction pile method · Design chart · Ultimate bearing capacity · Bearing capacity factor · Single column · Multiple columns

---

N. Aarthi (✉) · G. R. Dodagoudar
Geotechnical Engineering Division, Department of Civil Engg, IIT Madras, Chennai, Tamil Nadu, India
e-mail: aarthicivilian@gmail.com

B. Adimoolam et al. (eds.), *Deep Foundations for Infrastructure Development in India*, Lecture Notes in Civil Engineering 315, https://doi.org/10.1007/978-981-19-8598-0_5

Notation

ARR	Area replacement ratio
D	Diameter of the SCPs
EXP	Experiment
FEA	Finite Element Analysis
ICCF	Individual Column Composite Foundation
MCCF	Multiple Column Composite Foundation
S	Spacing between the SCPs
SCP	Sand Compaction Pile
UBC	Ultimate Bearing Capacity
d	Width/diameter of the foundation
γ	Insitu density of the soil-pre and post improvement
ϕ	Internal angle of friction of the pile material

1 Introduction

The stone column technique of ground improvement is a versatile method often adopted in the Western Hemisphere of the globe to attend issues related to geotechnical engineering. The inclusion of these granular columns not only improves the engineering properties of the composite ground but also accelerates the drainage to achieve the maximum settlement before the construction of the superstructure. One of the main reasons for the widespread implementation of the method is the availability of design methodologies (both empirical and non-empirical with simpler assumptions). These design procedures majorly include the framework in the form of charts, where one can find the unknown value (e.g., ultimate bearing capacity (UBC), stiffness improvement factor, β), with a few known factors (e.g., spacing and diameter of the columns, area replacement ratio adopted, internal friction angle of the columns, etc.). Promising notable design theories that are still used are proposed by Greenwood (1970), Hughes and Withers (1974), Hughes et al. (1975), and Balaam et al. (1977). Design methodology proposed by Priebe (1995) is still regarded as the most employed method to construct stone columns. Thus, it can be stated that design frameworks are abundant for adopting the stone column technique which resulted in its massive deployment worldwide.

Sand compaction piles (SCPs) are usually constructed by driving a casing pipe closed at the bottom with a special shoe through a loose to firm sand or a very soft to firm silt or clay stratum using a vibrator/ hammer located at the top of the casing. After driving, the casing pipe is filled with sand or a mixture of sand and aggregates in stages. The native sand is then densified by repeatedly raising and lowering the vibrating pipe as it is withdrawn from the ground. At this juncture, it can be clearly evident that these sorts of comprehensive design theories are found to be hardly any when the technique of sand compaction piles is taken into account and this forms the need for the present study. Upon reviewing the published data available on the sand compaction pile method, few studies limited to (D'Appolonia 1953; Webb and Hall 1969; Brown 1977) stated design methodologies in the form of charts. However, in these studies, the RD achieved was related to the dimensionless influence coefficients which were obtained by the extent of improvement achieved by the different types of vibroflot machines used for the installation of the SCPs. The major factor considered was the radial zone of improvement measured from the center of the SCP for a particular type of vibroflot machine employed in the construction, i.e., the design charts are specific to a particular type of vibroflot machine used. The Japanese geotechnical society (1998) has proposed a procedure using a pre-selected diameter of the SCP and target standard penetration test value (SPT-N) to design the SCP technique for a particular site. Kitazume (2005) has proposed a few design charts based on the accumulated field data obtained from the original SPT-N value of the native deposit and the target SPT-N value to be achieved for the site in order to design the improvement scheme using the SCPs.

The design charts proposed from the above past studies mentioned have considered the efficiency of vibroflot machines used for the construction of SCPs, and SPT-N from accumulated field data from sites. Thereby, these available charts do not apply to all field cases worldwide. Out of the few design frameworks available in the literature, as seen above, there are no proper solution methods available for the direct estimation of the ultimate bearing capacity (UBC) or axial capacity of the SCP improved sand deposits in the form of ready-to-use formulae and design charts. Moreover, a systematic framework involving the properties of the ground to be improved and the degree of improvement to be achieved, the size of the footing employed, and the characteristics of the SCPs to be installed are still lacking. Also, there are no existing generic design codes for sand compaction piles installed in sands unlike the other ground improvement techniques such as stone columns, dynamic compaction, etc. in India. Therefore it can be stated that even after a rough five-decade had passed since its invention, there is no commonly available design procedure available in the form of ready-to-use codes for computing SCP design techniques for a given site, unlike stone columns.

This forms the basis for the present study to develop the design charts using a few of the basic parameters of the sand deposit and SCPs for the direct estimation of the UBC of the SCP improved ground. The present attempt is an outcome of a detailed experimental (EXP) and finite element (FE) investigations conducted by Aarthi (2020). A series of plate load tests (using circular footing plate) were conducted (both in the laboratory (Fig. 1) and numerical simulation (Fig. 2)) on single SCP (Individual Column Composite Foundation—ICCF) and three SCP groups (Multiple Column Composite Foundation—MCCF) which include the compaction pile and their respective tributary areas too. The results obtained from the above tests are developed into a design framework from which one can directly estimate the ultimate bearing capacity of the SCP improved sandy deposits. The paper presents the results attained from the above tests in a nutshell in the form of design charts.

The input data needed to use these charts are the following: Initial relative density (RD) of the original ground to be treated, diameter (D) and spacing (S) of the SCP,

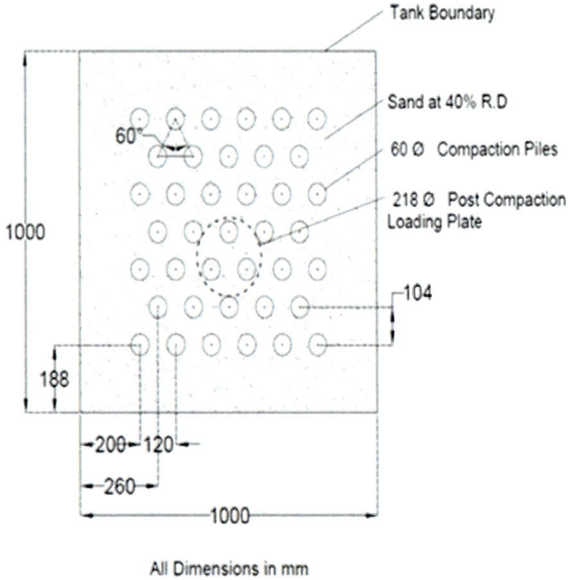

Fig. 1 Layout of SCP in test tank for experimental investigation in the laboratory (for 2D spacing)

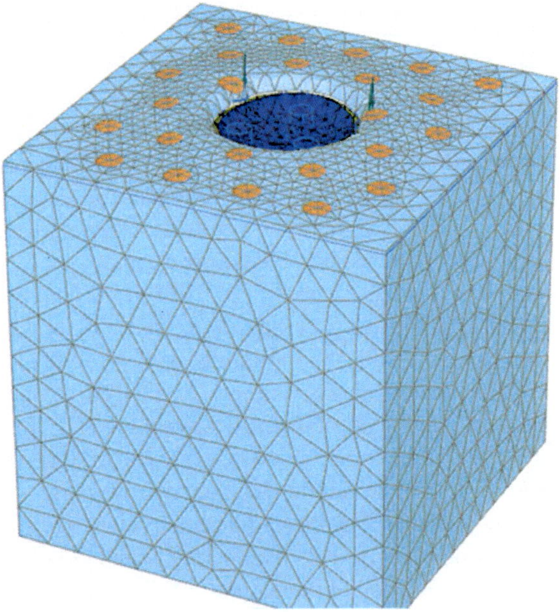

Fig. 2 View of the numerical model for lab-scale dimension showing failure after load application

target bulk unit weight of the sand deposit (γ) which is used for the assessment of the improvement, and diameter of the foundation (d). The output of the design chart is the UBC of the SCP improved sand deposit. The enhanced UBC of the SCP improved ground can be evaluated well in advance before the execution of the SCP treatment using the developed design charts. The design charts are developed for the equilateral triangular pattern of the SCPs installed in loose to medium dense sand deposits with fines content around 5%.

2 Factors Affecting the Design of SCP Treatment

The design of ground improvement using the SCP technique involves consideration of many factors that commonly affect the performance of the sand compaction piles for the proposed site. The factors are as follows:

- Insitu density of the soil—pre and post improvement (γ),
- Type of installation of the SCPs,
- Spacing between the SCPs (S),
- Diameter of the SCPs (D),
- Number of SCPs under loading (Single [ICCF] or Multiple [MCCF]),
- Type of installation pattern (Triangular or square or rectangular),
- Depth of improvement needed (L),
- Backfill material (pile material),
- Internal angle of friction of the pile material (ϕ),
- The withdrawal technique of the installation mechanism,
- Width/diameter (d) of the foundation,
- Area replacement ratio (ARR), and
- Vertical load intensity.

3 Design Charts for Axial Capacity of MCCF: Experimental Results

An extensive experimental study is carried out by Aarthi et al. (2019) to investigate the pressure-settlement response of the sand compaction pile improved cohesionless deposits. The results obtained from the above study are considered along with a comprehensive finite element (FE) studies performed by Aarthi (2020) using PLAXIS 3D, to aid in the development of design charts in the present paper. A humble beginning has been made to propose the design charts based on the extensive results of laboratory experiments and FE simulations performed on the single SCP (Individual Column Composite Foundation—ICCF) and SCP group (Multiple Column Composite Foundation—MCCF). Out of the many design charts developed the present section deals with the design chart developed using laboratory experiments presented in Aarthi et al. (2019). The following factors are considered to develop the design chart for the MCCF:

- Spacing and diameter of the SCPs,
- In situ density of the sand strata—post improvement, and
- Width or diameter of the foundation.

Table 1 Range of parameters employed in laboratory investigation

Parameter	Range
Initial relative density (RD) (%)	30, 40, 50, 60
Length of SCP (L)	500 mm (Leaving more than 4.5D below the bottom of the tank to avoid boundary effects)
Diameter of SCP (D)	600 mm
Spacing/Diameter (S/D) ratio	1.5, 2, 2.5, 3, 3.5
Size of the plate (d)	164, 218, 273, 328, and 382 mm For MCCF, from 1.5 to 3.5D spacing between the SCPs

The following two dimensionless numbers are related to each other in the design chart:

1. S/D ratio ⇒ Spacing of the SCP (m)/Diameter of the SCP (m).
2. Bearing capacity factor, $q_u/(\gamma * d)$ ⇒ Ultimate bearing capacity/(In situ unit weight of the improved deposit * diameter of the footing, i.e., loaded area).

In the above, q_u—in terms of kPa or kN/m^2, γ—in terms of kN/m^3, d—Diameter of the footing in m. The design chart gives the ultimate bearing capacity of the SCP treated sand deposit, for any desired value of the spacing between the SCPs, the diameter of the SCPs, size of the footing, and for the specified target unit weight required for the intended application and the given site conditions. Table 1 as presented below indicates the range of parameters that has been employed for the experimental investigation.

The design charts are developed for the equilateral triangular pattern of the sand compaction piles installed in loose to medium dense sand deposits with minimal fines content. The design chart as shown in Fig. 3 is established for the SCPs installed in sandbed by employing drop hammer method of installation and hence can be applied to similar executions. Owing to the economic viability of lab-scale experimentation, the drop hammer method of installation was chosen to drive the SCPs in the study. In the laboratory, the SCPs were installed by driving the casing in line with the desired pile configuration using a guiding rod and hammer. A hammer weight of 10 N was made to slide over a guide rod to drive the casing. The casing was driven with its bottom closed to prevent the entry of sand and the sand is fed through the casing to install the SCPs. Upon reaching the required depth, the hammer was removed and the calculated weight of sand was poured into the casing with the SCP occupying the inner diameter of the casing.

The installed SCP has to have the RD in the range of 70–75% for the effective utilization of the chart. The unit weight of the sandbed after improvement for the lab-scale experiment is obtained by back-calculating the relative density values by noting the penetration value using a self-weight penetrometer fabricated exclusively for facilitating the study. For field applications, the target relative density (RD) or the unit weight (γ) has to be fixed a priori before the execution of the SCP improvement technique. The RD or the unit weight of the unimproved ground can be ascertained by conducting the SPT, CPT, and geophysical tests also.

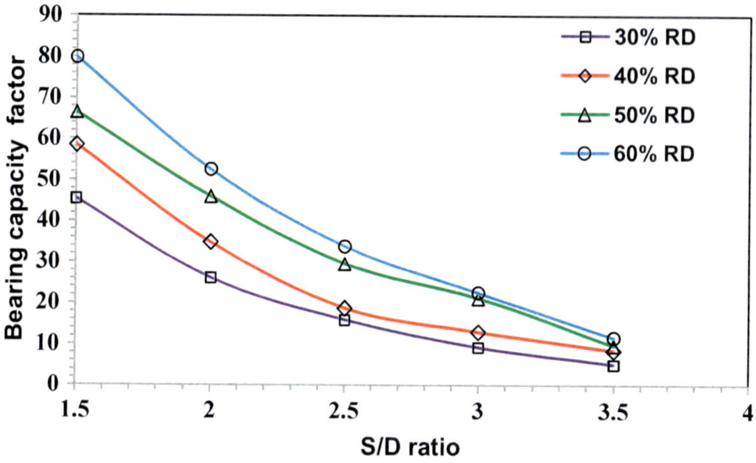

Fig. 3 Design chart for estimating the axial capacity of MCCF: Experimental results

4 Design Charts for Axial Capacity of MCCF: FE Results of Lab-Scale Model

Likewise in the previous section, a design chart is developed from the results of the FE simulation of the lab-scale SCP model. The FE results obtained from the preliminary sensitivity analyses are made use of in the development of the chart. In the parametric analyses, the initial RD of the sandbed is varied from 30 to 60% and the spacing between the SCPs is varied from 1.5 to 3.5D. The footing was supported by the three SCP groups (i.e., MCCF). The design chart will give a complete picture of the UBC or axial capacity attained from the FE results of the lab-scale SCP model as shown in Fig. 4. For the pre-selected S/D ratio of the SCP technique, size of the footing, and the required target unit weight to be achieved for the deposit, the design chart can be used to obtain the bearing capacity factor. Using the bearing capacity factor, the ultimate bearing capacity of the improved deposit can be evaluated w. r. t. the range of the initial RD (30 to 60%) of the sand deposits. The UBC values estimated from Fig. 4 are on average 12% less compared to those estimated from Fig. 3. Further refinements are needed in the finite element simulation studies related to the modeling of lab-scale SCP problem.

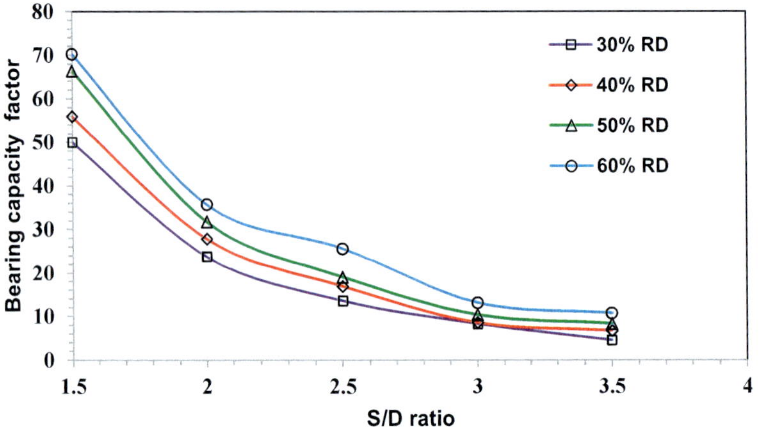

Fig. 4 Design chart for estimating the axial capacity of MCCF from FE results of the lab-scale model

5 Design Chart for Axial Capacity of ICCF: Experimental and FE Results

The FE simulation of the individual column composite foundation (ICCF) is also performed for the lab-scale model of the footing supported on a single SCP and its tributary area. The size of the footing is equal to effective diameter (d_e) = 1.05S, where S is the spacing between the SCPs. The experimental results were obtained for the ICCF installed in a sandbed with an initial RD of 40% where the spacing is varied from 2 to 3D. The experimental results are compared with the FE results and are used to develop the design chart as shown in Fig. 5.

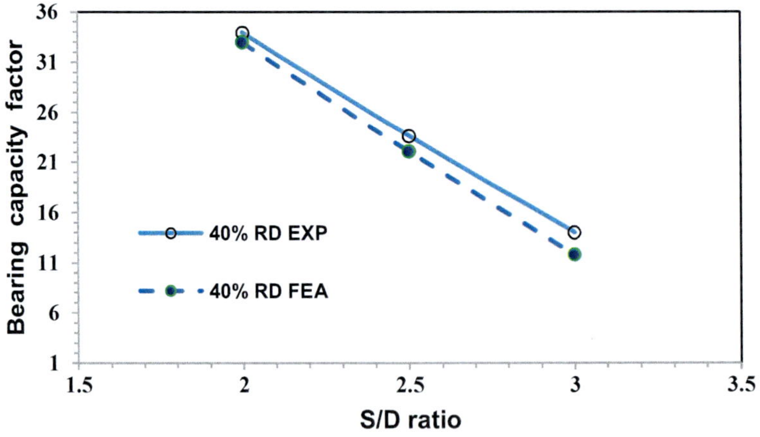

Fig. 5 Design chart for estimating the axial capacity of ICCF from experimental and FE results

As in the previous section, Fig. 5 also underestimates the axial capacity in the FE results, for the SCP treated ground for a given site condition in comparison to the experimental results. With the help of the design chart (Fig. 5), one can estimate the ultimate bearing capacity of the single SCP along with its tributary area loaded in the field, for the target unit weight of the deposit for which the improvement is proposed to be undertaken.

6 Design Chart for MCCF: Experimental and Field-Scale FE Results

In this section, an effort is made to combine the lab-scale experimental results and those obtained by the FE simulation of the field-scale SCP treated sand deposits to develop the design chart. Figure 6 gives the design chart for MCCF constructed in the sand deposit making use of the experimental and field-scale FE results. The SCPs are supposed to have a relative density in the range of 70–75% for effective utilization of the chart. The scale effect is evident as seen from the figure. A larger size laboratory experimental setup and rigorous field-scale experimentation are necessary to reduce the scale effects. Centrifuge experiments can also be performed to reduce the scale effects. However, in the present attempt, the results of scale-down laboratory experiments are used in the development of the design chart. The FE results obtained for the field-scale SCP problem need to be validated either using prototype or centrifuge results. Nevertheless, the design chart developed using the field-scale FE results can be used to make the preliminary designs for the SCP technique. In the FE simulation, the loaded footing comprises the three SCPs along with its tributary area, installed in the sand deposit with a layer of SCPs surrounding the footing as recommended by Han (2015).

Figure 6 also shows the regression equations developed using the experimental and FE results. The independent variable x in the equation designates the S/D ratio and y the *Bearing Capacity Factor*. These equations are applicable for the medium dense sand deposits. It can be stated that the regression equation developed for the estimation of the ultimate capacity of the MCCF using the FE results of the field-scale SCP problem is reliable. Similarly, a few more design charts can be developed by considering the different initial RDs for the field-scale problem of the SCP improved sand deposits. It is also evident from Fig. 6 that, the developed design framework, accommodates the results of both lab-scale experiments and field-scale experimental results obtained by finite element analyses.

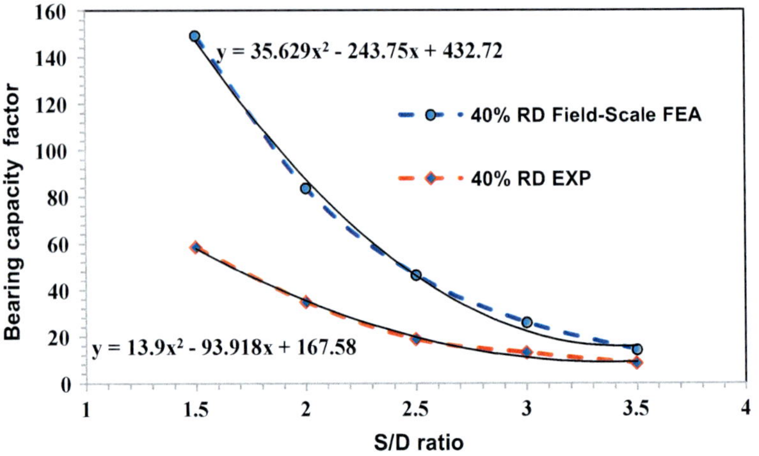

Fig. 6 Design chart for estimating the axial capacity of MCCF from lab-scale experimental results and FE results of the field-scale model

7 Conclusions

The paper presents a chart-based design framework based on lab-scale experimental work and FEA that needs to be validated and modified based on field-scale testing on cohesionless deposits improved by sand compaction piles. The design charts developed using laboratory experiments are found to have scale effects. Only a few basic parameters are needed to use the design charts to obtain the ultimate bearing capacity of the footing resting on a single sand compaction pile (SCP) and SCP group. These charts can be used to estimate the ultimate bearing capacity of the SCP treated sand deposit with pre-selected target unit weight of the treated deposit, diameter, and spacing of the SCPs to design the ground improvement scheme for the SCP method. The design chart developed using the results of finite element analysis of the field-scale SCP problem can be used to estimate the ultimate bearing capacity and accordingly the SCP technique be used to treat the cohesionless deposits in the field. More finite element analyses of the field-scale problem are further needed along with their validation for field-scale projects to develop more robust design charts that will be applicable to all conditions of the cohesionless deposits.

References

Aarthi N (2020) Experimental and numerical investigations of sand compaction piles installed in sandy deposits. Ph.D. Thesis. Indian Institute of Technology Madras, Department of Civil Engineering, Chennai, India

Aarthi N, Boominathan A, Gandhi SR (2019) Experimental study on the behaviour of sand compaction piles in sandy strata. Int J Geotech Eng. https://doi.org/10.1080/19386362.2019.1710391

Balaam NP, Poulos HG, Brown PT (1977) Settlement analysis of soft clays reinforced with granular piles. In: Proceedings of the 5th Asian conference on soil engineering, Bangkok. Thailand, 2–4 July, pp 81–92

Brown RE (1977) Vibroflotation compaction of cohesionless soils. Am Soc Civil Eng, ASCE J Geotechn Eng Div 103(GT12):1437–1451

D'Appolonia E (1953) Loose sands–their compaction by vibroflotation. In: Proceedings of Symposium on dynamic test soils, 56th annual meeting of American Society of Testing Materials, ASTM Special Technical Publication No. 156, Philadelphia, PA, pp 138–162

Greenwood DA (1970) Mechanical improvement of soils below ground surfaces. In: Proceedings of ground engineering, The Institution of Civil Engineers, London, United Kingdom, pp 11–22

Han J (2015) Principles and practice of ground improvement. John Wiley & Sons, New Jersey, pp 133–170

Hughes JMO, Withers NJ (1974) Reinforcing of soft cohesive soils with stone columns. Ground Eng 7(3):42–49

Hughes JMO, Withers NJ, Greenwood DA (1975) A field trial of the reinforcing effect of a stone column in soil. Geotechnique 25(1):31–44

Kitazume M (2005) The sand compaction pile method. Taylor & Francis, London, p 232

Priebe HJ (1995) The design of vibro replacement. Ground Eng 28(12):31–37

The Japanese geotechnical society (1998) Remedial measures against soil liquefaction: from investigation and design to implementation. A.A.Balkema Publishers, Rotterdam, Netherlands, pp 172–178

Webb DL, Hall RI (1969) Effects of vibroflotation on clayey sands. Am Soc Civil Eng, ASCE J Soil Mech Found Div 95:1365–1378

Ground Improvement Using Stone Columns to Mitigate Liquefaction, Reduce Settlements and to Increase Safe Bearing Capacity of In-Situ Soils—A Case Study

Venkata Krishnaiah Yeddala, Sai Vivek Adari, and Suresh Kumar Velugu

Abstract Assam cancer care foundation (ACCF) proposed the construction of hospitals and related facilities across 18 locations in the state of Assam-India. **Tezpur** is one among the 18 locations of Assam with high seismic risk. The strata at the site are comprised of mostly fine silty sands of lower SPT- N values. The water table is also encountered and ranging from 1 to 2 m below EGL. Since the site lies in seismic zone V as per IS 1893, detailed liquefaction analysis is carried out for all five boreholes and indicates that soil is liquefiable and ground improvement is inevitable up to a depth of 15 m below existing ground level. To mitigate the liquefaction potential and enhance the seismic performance of the soil, ground improvement using stone columns by Wet top feed Vibro replacement method is proposed. Also, the differential settlements can be avoided effectively under the raft foundation with Vibro Stone Columns. In this paper, field test validation results such as pre and post-SPTs and ECPTs to access the extent of ground improvement, along with other information adopted in this project presented.

Keywords Liquefaction · Stone column · SPT · ECPT · Wet top feed Vibro-replacement method

1 Introduction

M/s. L&T Constructions, Buildings & factories is awarded a contract by Assam Cancer Care foundation (ACCF) to construct hospitals and related facilities across 18 locations in the state of Assam. The 18 locations include expansion of State Cancer Institute of Assam. The entire state of Assam lies in the Seismic Zone-V and

V. K. Yeddala · S. V. Adari · S. K. Velugu (✉)
L&T Constructions, Chennai, India
e-mail: velugu@lntecc.com; velugusuresh@gmail.com

V. K. Yeddala
e-mail: yvkk@lntecc.com

S. V. Adari
e-mail: vivek-adari@lntecc.com

B. Adimoolam et al. (eds.), *Deep Foundations for Infrastructure Development in India*, Lecture Notes in Civil Engineering 315, https://doi.org/10.1007/978-981-19-8598-0_6

Fig. 1 Locations of Assam Cancer Care Hospitals

subjected to frequent earthquakes. **Tezpur** is one among the 18 locations of Assam at which, mainly the Hospital building (G + 3 Stories), LINAC block (G + 1) and service building and Ramp (G + 3 Stories) are to be constructed (Fig. 1).

Super structure's Structural and foundation system:

Buildings	No. of floors	Type of structural system
Hospital block	G + 3	RCC Beams and slabs with columns/walls with base isolators
LINAC block	G + 1	Same structural system, but w/o base isolators
Service Building	G + 3	Same structural system, but w/o base isolators

The minimum safe bearing capacity of 20 T/m^2 is required for superstructure at depth of 3 m below EGL (Existing Ground Level).

2 Site Description and Sub-Soil Profile

The proposed site is a low-lying area with a water table close to the ground surface as the Brahmaputra river is close by. To assess the subsoil conditions, the soil exploration is carried out all over the site up to a maximum depth of 20 m. It is observed that the site is underlain by top 1 m with filled up soil followed by silty sand ($N < 10$) up to 4.5 m below EGL. This layer is underlain by silty fine to medium dense sand ($N < 20$) up to 12 m followed by dense silty sand with mica ($N > 25$) up to the final exploration depth. At the time of soil investigation, the groundwater table is observed between 1.0 and 2.0 m below the natural ground level. The generalized soil profile

Table 1 Generalized soil profile

S. No	Soil strata	Depth range below NGL(m)	Field SPT-N value
1	Filled up soil	0.0–1.0	–
2	SW-SM	1.0–4.5	6
3	SP	4.5–9.0	13
4	SW-SM	9.0–12.0	18
5	SP	12.0–16.5	26
6	SW-SM	16.5–20.0	39

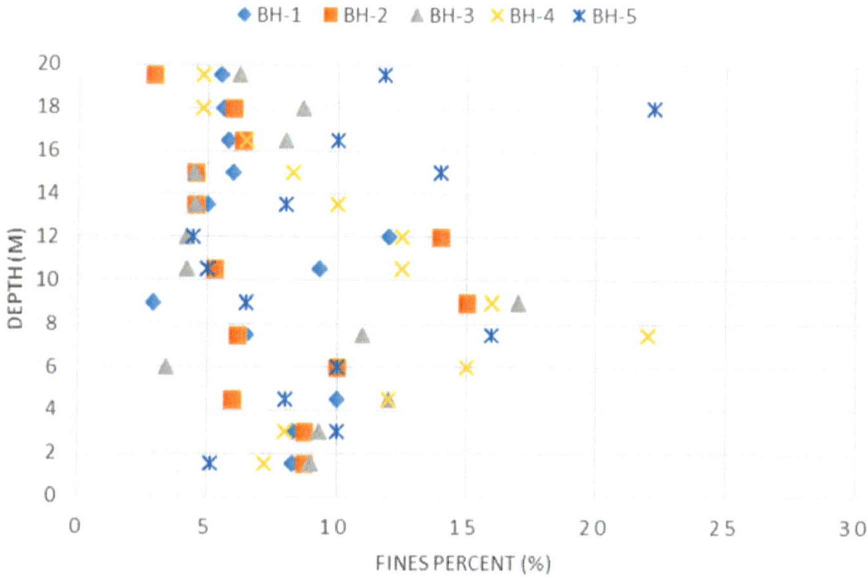

Fig. 2 Percentage of fines versus depth

based on bore-logs for design and analysis is shown in Table 1. Figure 2 represents the percentage of fines passing through 75 μm sieve with depth.

3 Seismic Performance of the Ground

Because of the presence of water table almost at ground level and fine silty sand layers till 16 m depth, the susceptibility of liquefaction is checked using Seed and Idris's method (Seed and Idriss 1971, Youd and Idriss 2001). The predicted Earthquake magnitude is 7 and peak ground acceleration (PGA) is 0.36 g. At the time of soil investigation, the groundwater table is observed between 1.0 and 2.0 m below the ground level, however, for design, the groundwater table is considered at the ground surface for all the calculations based on the peak rainfall season as the site is close to Brahmaputra River. The liquefaction check is carried out for all five boreholes and liquefaction potential is observed up to 13–15 m depth below the Existing Ground Level. The factor of safety against liquefaction for the site is less than 1 and hence it will liquefy for the considered ground motion and induces settlements.

The elastic immediate settlement of the raft due to static loads is calculated based on the theory of Elasticity (Bowles 2012). Ground deformation due to the seismically induced liquefaction is also calculated based on the Post-Liquefaction Volumetric Strain (Ishihara and Yoshimine 1992).

For the raft foundation with the load intensity of 20 T/m^2, the estimated static and seismic settlements exceeded permissible settlements of 75 mm as per IS 8009 code and hence ground improvement is required.

4 Ground Improvement Using Vibro-Replacement

Several methods are considered to mitigate the liquefaction potential by improving the resistance of the soils, to provide the required minimum bearing capacity and to restrict total, differential settlements to acceptable magnitudes under the proposed structural loading.

However, the selection of the suitable ground improvement technique is governed by various factors such as the fines content of the soil, degree of saturation, availability of space for equipment, environmental regulations, and cost factor. Based on the evaluation of these, Vibro-Replacement method is considered as the most suitable technique of ground improvement. Because in general, Vibro-Compaction method or Deep Dynamic Compaction is preferable only if the entire soil is sandy but it is uneconomical to use in case Silt Content is more than 15% by weight or 2% Clay even though Vibro-Compaction and Vibro-Replacement construction techniques are almost similar in the procedure.

Due to vibration during the installation of Stone Columns, this method increases the density of the soil, thereby increasing the cyclic shear resistance. It also reduces the Cyclic Shear Stress and provides the drainage due to inclusion of Stone aggregates which in turn mitigates the liquefaction potential.

Design Challenges

Non-availability of Clear guidelines: No IS/BS/Euro code is available to determine the grid spacing and diameter of Stone Columns in non-homogenous soils/Sandy Soils.

Establishment of validation of Ground Improvement: Validation of Ground Improvement is also a major issue. Type of Validation is established based on the type of soil strata. The validation is done to check the mitigation of liquefaction and load-bearing capacity of the improved ground.

Despite all these, Vibro Stone Columns of 0.9 m diameter with a spacing of 2 m c/c with the square pattern are proposed all over the building footprint area despite no

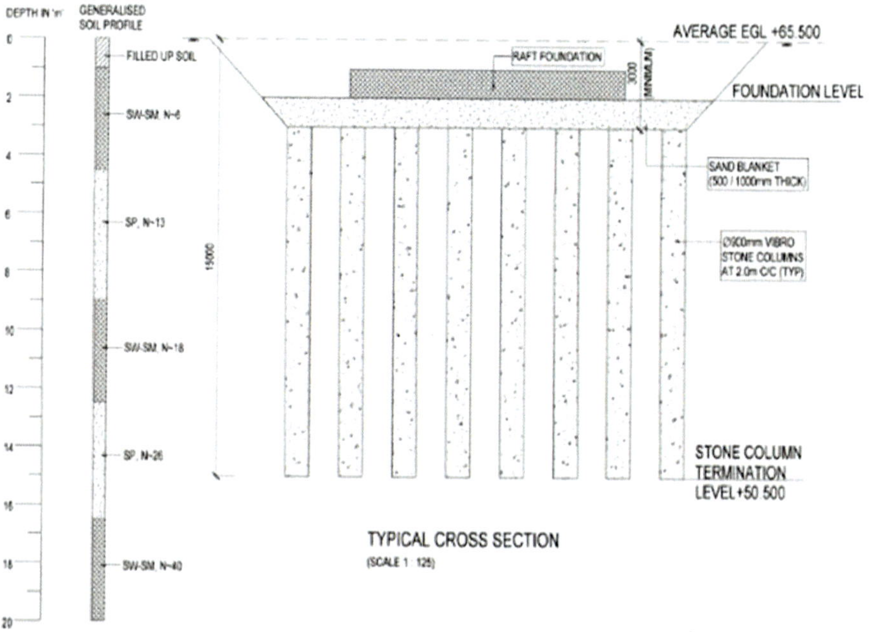

Fig. 3 Sketch showing the typical arrangement of the stone column below Raft

availability of IS code. The spacing of the stone columns is found out by designing like as a sand compaction pile (ref: the sand compaction pile method by Masaki Kitazume 2005) in the iteration process. Priebe method is used to calculate the settlements of the improved ground and to check the liquefaction after the execution of stone columns. A sand blanket layer of 500 mm is provided and compacted in layers to a relative density of 80% for proper load distribution and to act as a drainage layer for pore water pressure dissipation.

The typical arrangement of the stone column below the Raft footing is shown in Fig. 3. The minimum depth of the Footing proposed is 2.5 m below the leveled Existing Ground Level (EGL) to provide the confinement to the footing.

Stone column design parameters

The angle of internal friction = 40°

Unit weight of column material = 22 kN/m^3.

5 Installation of Stone Columns

Wet Vibro-Replacement method is proposed instead of Dry Vibro Displacement method since the stone columns are extending up to 15 m length below NGL and at the depths of 12 to 15 m, SPT is greater than 20, hence it is difficult to install stone columns by Dry Vibro displacement method. While installing stone columns by Wet Vibro top feed method, a huge amount of muck is generated and its disposal is a major issue. However, precautions are taken to maintain the site good conditions, i.e., proper drainage facilities are adopted at the site (Fig. 4). A total of 998 stone columns are proposed over a footprint area of 3835 m^2. One to two additional rows of stone columns is provided outside the raft footprint area to provide confinement. A photograph of stone columns after installation at Tezpur site is presented in Fig. 5.

Stone Consumption:

The consumption of stone backfill shall not be less than the theoretical volume of the column (calculated based on the column diameter specified) plus additional volume due to compaction effort and losses during installation. The rate of stone consumption shall be assessed by averaging the material used in 10 adjacent compaction points. As per our records, stone consumption is 12–20% more than the theoretical volume. Table 2 indicates the stone column aggregate consumption.

Fig. 4 Typical crane setup by Wet Top Feed Method at Tezpur site (Courtesy: Keller)

Table 2 The Stone column aggregate consumption

Material reconciliation					
S.no	No. of days	No of stone column achieved	Theoretical aggregate consumption (Cum)	Actual aggregate consumption (Cum)	Difference
1	28	998	7557.91	8542.12	12.8%

6 Quality Assurance/Quality Control of Stone Columns

Quality Check for Stone Aggregates

As a minimum, the following controls are adopted to maintain a good quality of stones for the Vibro stone column works:

- The aggregates should be chemically inert, hard, and resistant to breakage.
- Well graded gravel with gradation ranging from 50 to 12 mm is used for Vibro-Replacement stone columns. The gradation is chosen in such a way the voids between the stones are filled by the in-situ material which is suspended in water to form a rigid column. The gradation of stone aggregates is given in Table 3.

Aggregate quality requirement for Vibro stone column work is given below in Table 4.

The stone columns are terminated at a depth of 15 m below the natural ground level depending on the penetration resistance achieved through the vibrator. The degree of compaction is measured in terms of electric current amperage. During Vibro-compaction, quality control can be implemented to ensure Vibro-compaction

Fig. 5 Stone columns after installation at Tezpur site

Table 3 Gradation for vibro-replacement stone columns

Sieve size (mm)	% Passing, By weight
50	100
37.5	90–100
25	30–80
20	10–50
14	0–2

Table 4 Quality Requirements for Stone column aggregates

Type of tests	Code	Criteria
Crushing value	IS 383	<30%
Impact value	IS 383	<30%
Stone gradation	IS 383	75 to 12 mm

to achieve desired performance through amperage measurement of the vibrator. The quality control items include the location of compaction points, depth of penetration, vibratory power consumption or electric current amperage, frequency and duration of penetration, compaction, and extraction, depth of obstruction encountered if any, and ground subsidence, all these are measured through M4 graphs which are generated automatically during the installation. As mentioned earlier, water jetting is used to create 900 mm diameter stone columns. It is important to have well-planned water flow channels to direct water from compaction points into settling ponds.

Execution of Vibro stone column works requires quality which is assured by implementing various measures at different stages.

These are as follows:

- During construction—Monitoring and recording of process parameters
- After construction—Post Treatment Soil Investigation using SPT and ECPTs.

Automated Real-Time Monitoring:

The monitoring of each stone column is performed using an automated computerized recording device. This instrument yields a computer record of the installation process in a continuous graphical mode plotting depth versus time and power consumption (compaction effort) versus time. The information provided includes:

- Stone column reference number
- Date of installation
- The period required for installation
- Maximum depth
- Compaction effort during penetration and compaction process.

These graphs are the main quality control tools during the installation process. The M4 graph of stone column no.680 which is executed at the site can be shown in Fig. 6.

7 Validation of Ground Improvement

To assess the extent of improvement, field tests such as Standard penetration tests (SPT) and Electrical cone penetration tests (ECPT) are carried out at the site since these are the best tests for the evaluation of seismic liquefaction potential. After

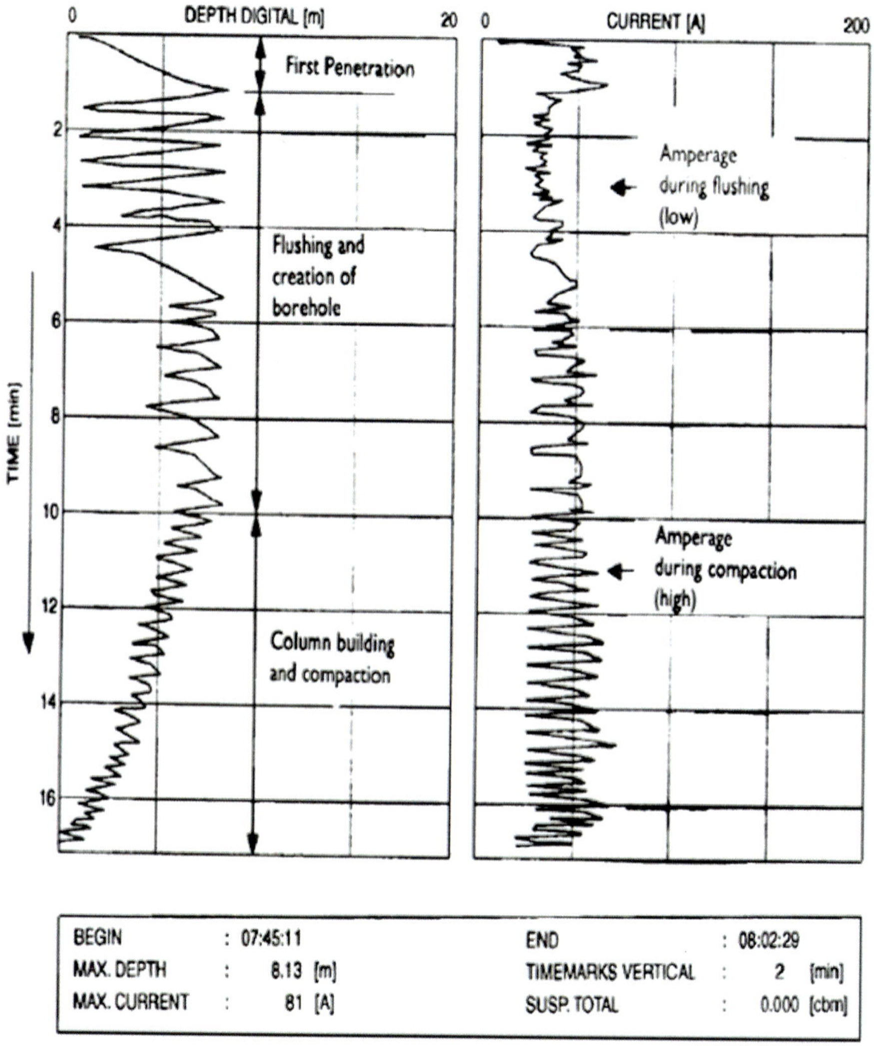

Fig. 6 M4 graph generated during the installation of the stone column

installing stone columns by Wet Vibro top feed replacement method. 4 No's of Standard penetration tests (SPT) and 4 No's of Electrical cone penetration tests (ECPT) are carried out at specific locations to provide a continuous record of penetration resistance of treated ground between stone columns. Figure 7 shows pre and post SPT test plot.

From Fig 7, it is observed that the soil which is in loose condition with SPT N value in the range of 2 to 10 is densified to 2 to 3.5 times of initial SPT values, and the soil in the medium dense condition is improved by 1.3 to 2 times. Figure 8 indicates the improvement of CRR after the installation stone columns.

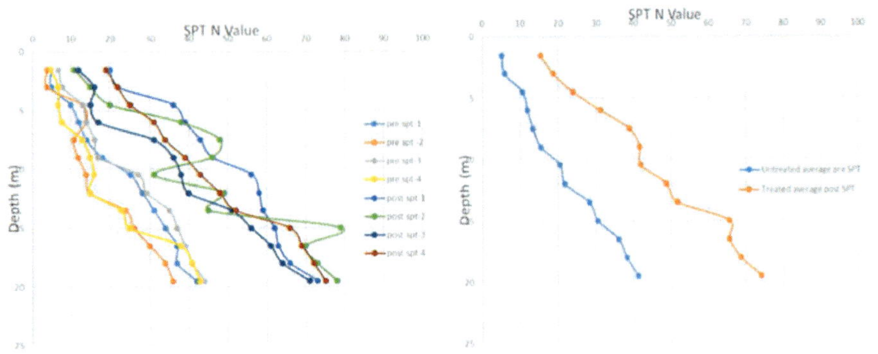

Fig. 7 SPT N plot before and after treatment

Fig. 8 Variation in cyclic resistance ratios (CRR) before and after treatment

As mentioned earlier, to validate the ground improvement pre and post the ground improvement, ECPTs are conducted. ECPT readings are taken for every 20 cm interval. Figure 9 shows the Cone tip resistance of pre and post ECPT tests; Fig. 10 shows the Sleeve friction plot of pre and post ECPT tests. Figures 11 and 12 present the relative density profiles before and after the ground improvement and the factor of safety against liquefaction.

Based on post penetration tests results such as SPT, ECPT, post liquefaction analysis are done using Seed and Idriss Method and observed that the ground improvement has shown significant results against liquefaction. The analysis indicated that the resistance of the soil is improved and it is non-liquefiable and hence the design intent is achieved.

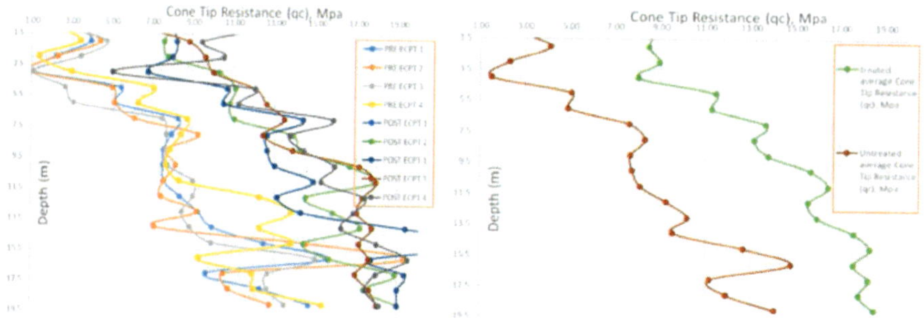

Fig. 9 Cone tip resistance plots from pre and post ECPT tests

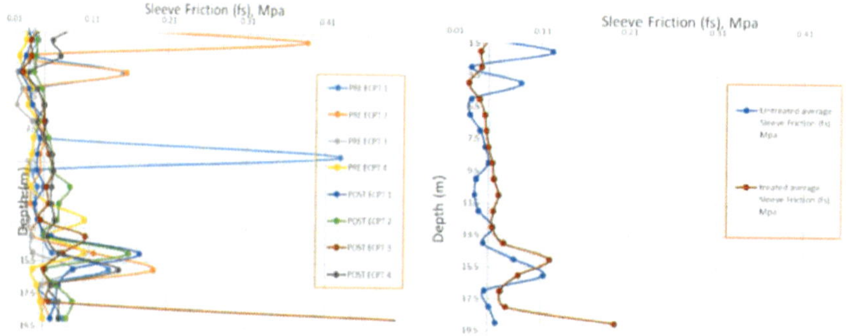

Fig. 10 Sleeve friction plots from pre and post ECPT tests

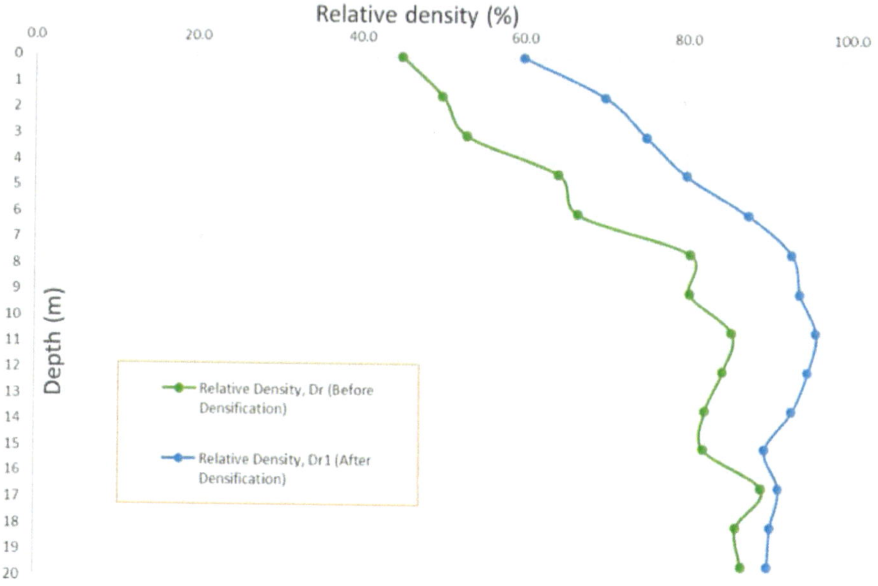

Fig. 11 Relative density plot before and after ground improvement (estimated from cone resistance)

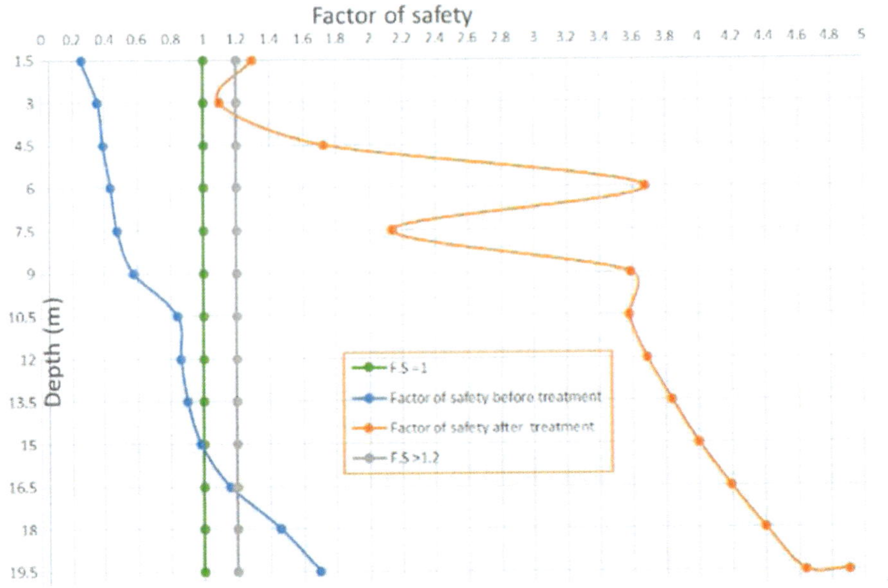

Fig. 12 The factor of safety against liquefaction before and after ground improvement

8 Conclusion

In the study detailed in this paper, the designers recommended the ground improvement technique using stone columns as an effective method for mitigating the liquefaction potential of saturated fine silty sands. For the same project site, the suitability of pile foundation is also checked but discarded as no lateral resistance would be offered by piles during an earthquake considering liquefaction prone layers at shallow depths and extending up to an average depth of 15 m. As mentioned above replacement ratio of 16% considered, however, fines content varies throughout site, it is better to consider 15–20% of area replacement ratio for better results when wet top feed method is adopted. Also, it is recommended to conduct Post ground improvement validation tests such as SPT, ECPT after 7–14 days of installation of stone columns. The muck generation will be an issue at sites, proper drainage facilities must be taken care of. It is also observed that the cost of aggregates impacts the cost of the project so, proper variation in the quantity of the aggregates must be considered.

Acknowledgements Authors also thankful to **Dr. Subhadeep Banerjee,** Associate Professor of Geotechnical Engineering, Indian Institute of Technology Madras, Chennai for his support and suggestions. He wholeheartedly gave appointments for our clarifications and doubts despite his busy schedules.

The authors acknowledge the help rendered by the geotechnical team of **M/s. Keller ground engineering Pvt. Ltd** for the successful execution of stone columns and conduction of post field tests such as SPT and ECPT.

References

Bowles JE (2012) Foundation analysis and design, 5th edn

IS 1893 (Part-1) (2016) Criteria for earthquake resistant design of structures–part 1: general provisions and buildings. Bureau of Indian Standards, New Delhi

Ishihara K, Yoshimine M (1992) Evaluation of settlements in sand deposits following liquefaction during earthquakes

Kitazume M (2005) The sand compaction pile method. Taylor and Francis

Priebe HJ (1998) Vibro replacement to prevent earthquake-induced liquefaction. Proc Geotechn Colloquium at Darmstadt, Germany. Technical paper 12-57E

Seed HB, Idriss IM (1971) Simplified procedure for evaluating soil liquefaction potential. J Geotech Eng Div, ASCE 97(9):1249–1273

Youd TL, Idriss IM (2001) Liquefaction resistance of soils: summary report from the 1996 NCEER and 1998 NCEER/NSF workshops on evaluation of liquefaction resistance of soils. J Geotech Geo Environ Eng ASCE 127(4):297–313

Piling and Deep Foundation Techniques

A Study on the Evaluation of Pile Bearing Capacity Factor and Adhesion Factor in IS 2911

Ashirbad Satapathy, Ramanand Shukla, Sanket Rawat, and Ravi Kant Mittal

Abstract Bearing capacity factor (N_q) and adhesion factor (α) are the key parameters for analysing the load carrying capacity of pile foundations embedded in cohesionless and cohesive soils, respectively. Numerous models are available in the form of charts or equations for the computation of these parameters and, hence, have been adopted by most of the international and national standards. Indian Standard for the Design and Construction of Concrete Pile Foundations, IS 2911 Part 1 (2010), also makes use of charts detailed in its Annexure B. However, the sources of these charts have not been specified clearly, making it difficult to back-refer to the actual models to assess their basis for resolving any critical scenarios encountered during design. The present study aims at acquiring a distinct understanding of the development of these charts in order to bring more clarity to the design process of pile foundations. The given charts have been compared with various models specified in the existing literature and international standards for the calculation of N_q and α for both driven and bored piles and, hence, the basis of Indian Standard charts has been identified. Moreover, it is evident that the use of these charts complicates the process of automation of analysis and design of pile foundations as well as the associated optimisation studies. Additionally, the manual entry of data especially, from the logarithmic graph of N_q, escalates the chances of error thus, making it a critical concern for design offices. Therefore, this paper also presents a non-linear regression model for N_q and α, developed using NCSS software for both driven and bored piles. Through multiple iterations, the value of coefficient of determination for N_q and α has been found to reach greater than 0.995. The developed equations can be simplistically used for both manual and automated analysis of pile foundations.

Keywords Adhesion factor · Bearing capacity factor · Pile foundation · Regression equations

A. Satapathy (✉) · R. Shukla · S. Rawat · R. K. Mittal
BITS-Pilani, Pilani Campus, Jhunjhunu, Rajasthan, India
e-mail: satapathy.shrbd@gmail.com

R. K. Mittal
e-mail: ravimittal@pilani.bits-pilani.ac.in

B. Adimoolam et al. (eds.), *Deep Foundations for Infrastructure Development in India*, Lecture Notes in Civil Engineering 315, https://doi.org/10.1007/978-981-19-8598-0_7

1 Introduction

The requirement for the installation of pile foundations arises primarily due to low bearing capacity of the soil at the construction site. Use of pile foundations at such ground conditions alleviates the risk of excessive settlement. Even though pile foundations are widely adopted for the design of numerous structures, accurate evaluation of the ultimate load carrying capacity of pile foundations still remains a challenging task for engineers due to the uncertainties in the soil profile and the absence of a detailed methodology for the same. To reduce the complexity of the analysis, several researchers have suggested a simplified methodology for the assessment of load carrying capacity of pile foundations. The associated guidelines have also been prescribed in various national and international standards such as BS EN 1997 Part 1 (2004), FHWA NHI (2016), IS 2911 Part 1: Sects. 1–4 (2010a, b, c, d), and API 2A-WSD (2007). The design of foundations for any structure involves the computation of the required dimensions which depends upon the bearing capacity of the soil. Therefore, the accurate evaluation of bearing capacity of the soil from field soil investigation data or laboratory testing is of crucial importance in design. The assessment of bearing capacity of cohesionless and cohesive soils is greatly affected by the calculation of bearing capacity factor, N_q, and adhesion factor, α, respectively. Several researchers have proposed different values and equations for N_q with angle of internal friction, ϕ, as the variable. In case of cohesive soils, adhesion factor, α, is needed for the calculation of shaft resistance of a pile which varies with the undrained strength of the soil. Interestingly, it can be noted that the bearing capacity obtained from different methods may not always yield identical results and this variation primarily arises due to the inconsistent procedures for the calculation of N_q and α.

The present study is mainly focused on the procedure for computation of N_q and α as specified in the Indian Standard for Design and Construction of Concrete Pile Foundations, IS 2911 Part 1: Sect.1–4 (2010a, b, c, d) and aims to provide a better understanding of the same by the identification of the source of these charts and comparison of the results with the models suggested by various researchers. A non-linear regression analysis has been implemented in order to provide simple equations for the evaluation of N_q and α, thus making an attempt towards streamlining the automation of analysis procedures of pile foundations. The following sections briefly highlight the various existing procedures for the computation of N_q and α.

2 Existing Methods for Prediction of N_q and α

Various researchers have proposed different methodologies for the computation of bearing capacity factors. Amongst them, some are widely used and practiced for the design of pile foundations in field applications. This paper focuses on some of these methods for study and their comparison with that of IS 2911 Part 1: Sect.1–4 (2010a, b, c, d) procedure.

Existing Methods for Nq prediction

The values of N_q may differ depending upon the type of installation, i.e. driven pile and bored pile. For driven pile foundation, works of Terzaghi (1943), De Beer (1945), Brinch Hansen (1951, 1961), Meyerhof (1953), Skempton (1951), Caquot and Kerisel (1956), Berezantsev et al. (1961) and Vesic (1963) have been studied with author's focus limited to the bearing capacity factors. Figures 1 and 2 show the plot of N_q given by these authors and their comparison with Indian code values for driven and bored piles, respectively. The differences in the N_q values of different methods can clearly be visualised from the plots. For bored piles, works of Meyerhof (1953), Thurman (1964), Janbu (1976), Vesic (1977), etc., are considered. Janbu (1976) provided a relation between N_q and ϕ which can easily be used to work out N_q as specified in Eq. 1.

$$N_q = \left[\tan\phi + \sqrt{1 + \tan\phi^2}\right]^2 e^{2\theta\tan\phi} \quad (1)$$

where N_q is the pile bearing capacity factor from Fig. 2 and ϕ is the angle of internal friction. θ is the angle defining the shape of shear surface around the tip of the pile. It varies between $\pi/3$ and $0.58\ \pi$.

Thurman (1964) proposed a method to calculate pile tip resistance as given in Eq. 2.

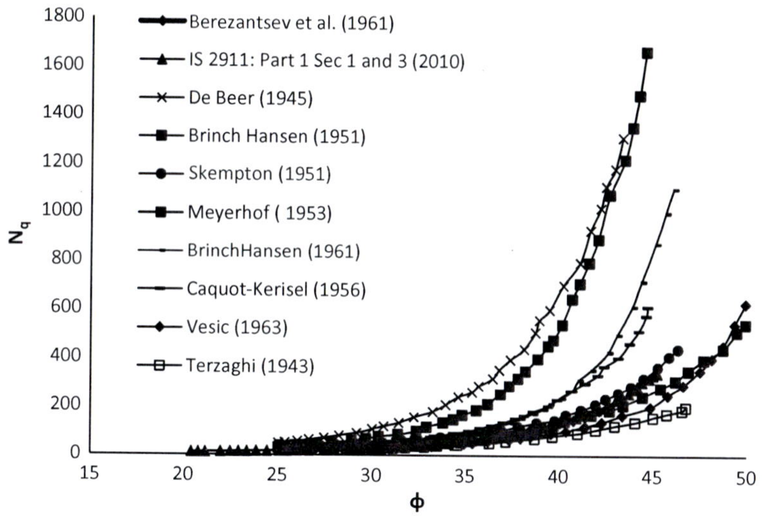

Fig. 1 N_q versus ϕ Comparison with IS 2911 part 1: Sects. 1 and 3 (2010a, c)

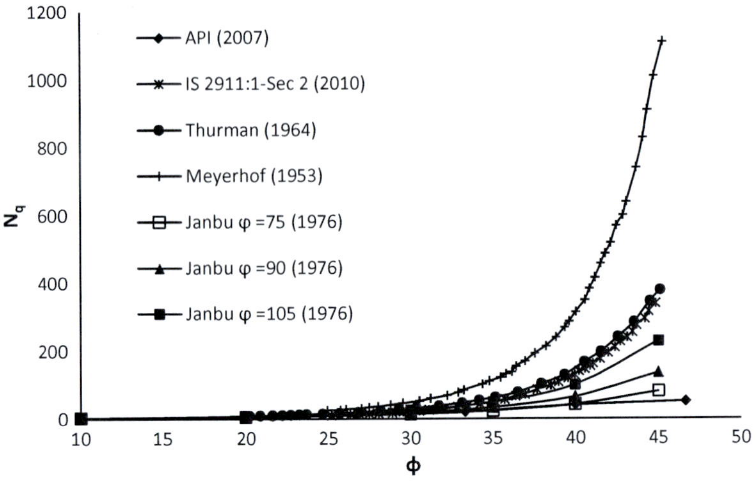

Fig. 2 N_q versus ϕ Comparison with IS 2911 part 1: Sects. 2 and 4 (2010b, d)

$$Q_p = A_p \alpha_T \sigma^{'}_{vt} N_q \tag{2}$$

where Q_p = pile tip resistance, A_p = area of pile tip, α_T = dimensionless factor provided by Thurman (1964) in the form of chart, $\sigma^{'}_{vt}$ = effective vertical stress at pile tip and N_q = bearing capacity factor from Fig. 2.

Vesic (1963) suggested that the base resistance of pile is not governed by vertical ground pressure but by the mean effective normal ground stress, σ_m. The bearing capacity is given by Eq. 3.

$$q_b = cN_c^* + \sigma_m N_\sigma^* \tag{3}$$

where q_b = bearing capacity of soil, c = unit cohesion, σ_m = mean effective normal ground stress,

$q_o^{'}$ = effective vertical pressure at the base level of the pile and $N_c{}^*$ and $N_\sigma{}^*$ = bearing capacity factors related to each other by the Eqs. 4–6.

$$\sigma_m = \frac{1 + 2K_o}{3} q_o^{'} \tag{4}$$

$$N_\sigma^* = \frac{3N_q^*}{1 + 2K_o} \tag{5}$$

$$N_c^* = \left(N_q^* - 1\right)\cot\phi \tag{6}$$

where K_0 = coefficient of lateral earth pressure at rest.

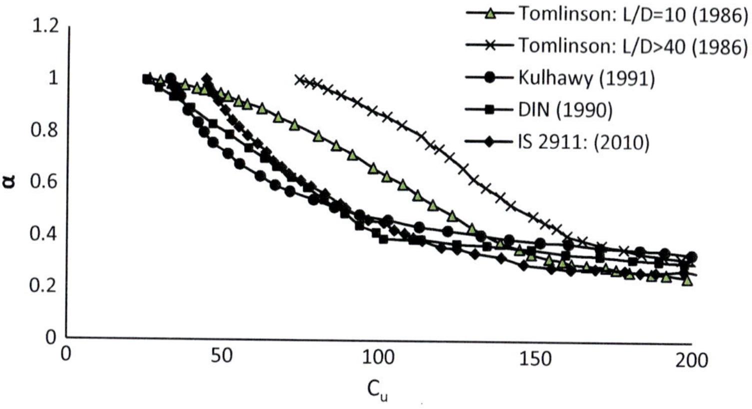

Fig. 3 Comparison of α in IS 2911 Part 1: Sect.1–4 (2010a, b, c, d) with various sources

Existing Methods for Adhesion Factor (α)

Indian code provides the value of adhesion factor, α, in the form of a plot. Various models of variation of adhesion factor, α, with C_u (undrained cohesion) are proposed by researchers and standards such as Tomlinson (1986) for L/D = 10 and L/D > 40, Kulhawy (1991), etc., as shown in Fig. 3.

3 Provisions Available in National and International Standards

AASHTO (2017)

Section 10 of AASHTO (2017) mentions guidelines for pile foundations and have provided a plot for bearing capacity factor, N_q, varying with soil friction angle, ϕ, to calculate tip resistance using Nordlund (1963)/Thurman (1964) method as shown in Fig. 4.

The procedure for the computation of adhesion factor α, as shown in Fig. 5, is adopted depending upon the soil profile and is given after Tomlinson (1986). This adhesion factor is used to calculate the shaft resistance of pile foundation and this method is known as the α-method.

BS 8004 (2015)

The British Standard, BS 8004 (2015), suggests the use of values of N_q given by Berezantsev et al. (1961) as in Fig. 1 for the calculation of base resistance.

GEO 1/96: Foundation Design and Construction, (2016)

The Geotechnical Engineering Office of the Government of the Hong Kong Special Administrative Region published a reference document for pile design, Foundation

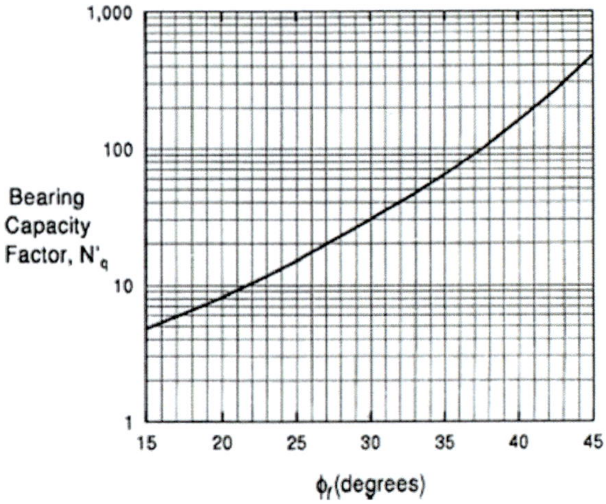

Fig. 4 Bearing Capacity Factor, Nq (AASHTO 2017)

Design and Construction, GEO Publication No. 1/96, have provided the values of N_q taken from Poulos and Davis (1980) as shown in Fig. 6.

API 2A-WSD (2007)

Guidelines in the American Petroleum Institute API 2A-WSD (2007) have recommended the values of N_q, varying with δ (soil–pile friction angle), where $\delta = 0.75$ ϕ, as shown in Table 1.

FEMA : Coastal construction manual (2011)

FEMA, in its 4th edition of the Coastal Construction Manual (2011), has provided the values of bearing capacity factors corresponding to various values of ϕ as shown in Table 2.

DIN 4014 (1990)

The German Institute of Standardization (DIN) in its standard DIN 4014 (1990) for bored cast-in-place piles has provided values of adhesion factor, α, in the form of a graph as shown in Fig. 3. The values of α are very similar to that provided in the Indian Standard IS 2911 part 1 (2010).

FHWA NHI, Geotechnical Engineering Circular no. 012 (2016)

The Federal Highway Administration (FHWA) in its NHI Courses No. 132,021 and 132,022, Design and Construction of Driven Pile Foundations—Volume I, makes use of the values of N_q provided by Bowles (1977) for calculation of pile resistance in cohesionless soil for which the bearing capacity factors and adhesion factors are presented in the form of plots as shown in Fig. 7.

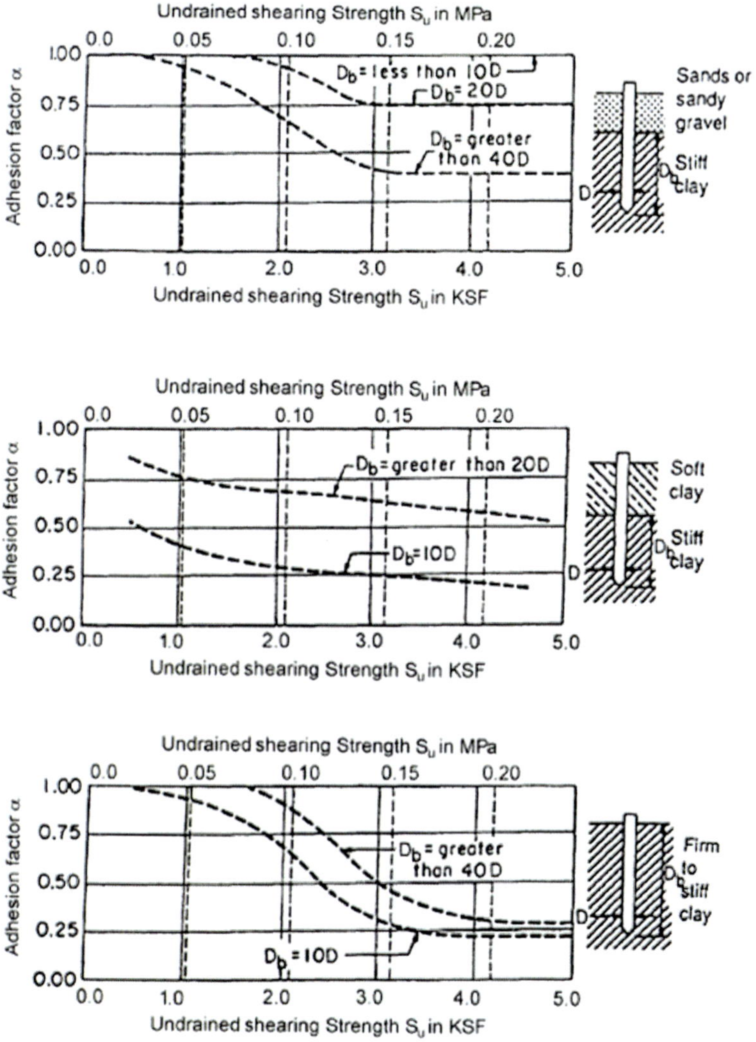

Fig. 5 Adhesion factor, α varying with soil profile (AASHTO 2017)

IS 2911 Part 1: Sect.1–4 (2010a, b, c, d)

The Indian code for Design and Construction of Pile foundation IS 2911 Part 1 in its Sects. 1–4 has provided plots of bearing capacity factor, N_q, and adhesion factor, α, for both bored and driven piles, and these plots are subsequently meant to be used in the design or analysis of pile foundations. The plots from the codes are shown below in Figs. 8 and 9.

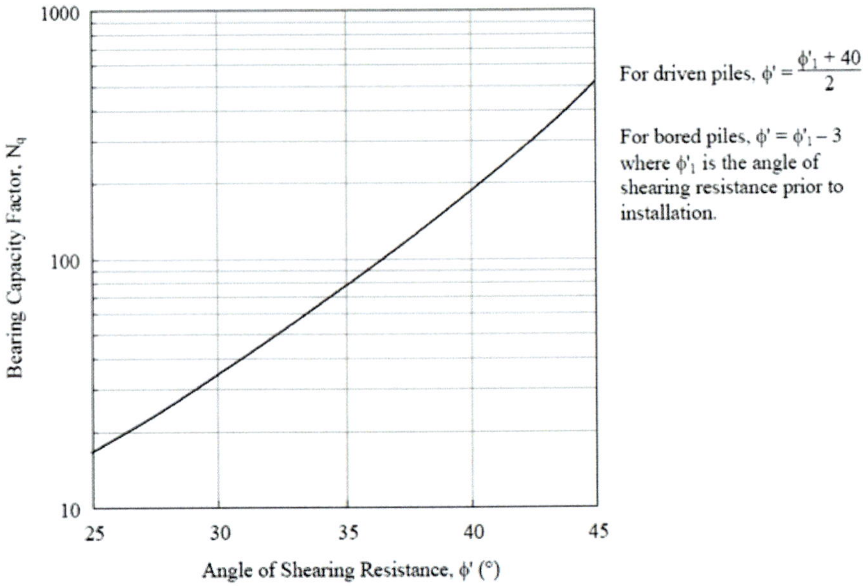

Fig. 6 Relationship between N_q and ϕ' (Poulos and Davis 1980)

Table 1 Values of bearing capacity factor recommended by API 2A-WSD (2007)

δ (degrees)	15	20	25	30	35
N_q	8	12	20	40	50

Table 2 Values of bearing capacity factor recommended by FEMA (2011)

ϕ (degrees)	26	28	30	31	32	33	34	35	36	37	38	39	40
N_q (driven)	10	15	21	24	29	34	42	50	62	77	86	120	145
N_q (bored)	5	8	10	12	14	17	21	25	30	38	43	60	72

4 Comparison of IS 2911 Factors With Existing Methods and Codes

The values/plot of N_q and α, provided in the Indian code for design of pile foundations in its Sects. 1–4, are compared with the values provided by different researchers and international codes with an objective to conclude the possible source of the values in the Indian code. As obtaining exact values from the graphs and their comparison was not possible, Java-based Plot Digitizer software was used to digitise the plots and the obtained coordinates were plotted separately in Microsoft Excel. Obtained graphs were compared with those given in Indian codes as shown in Figs. 1, 2 and 3.

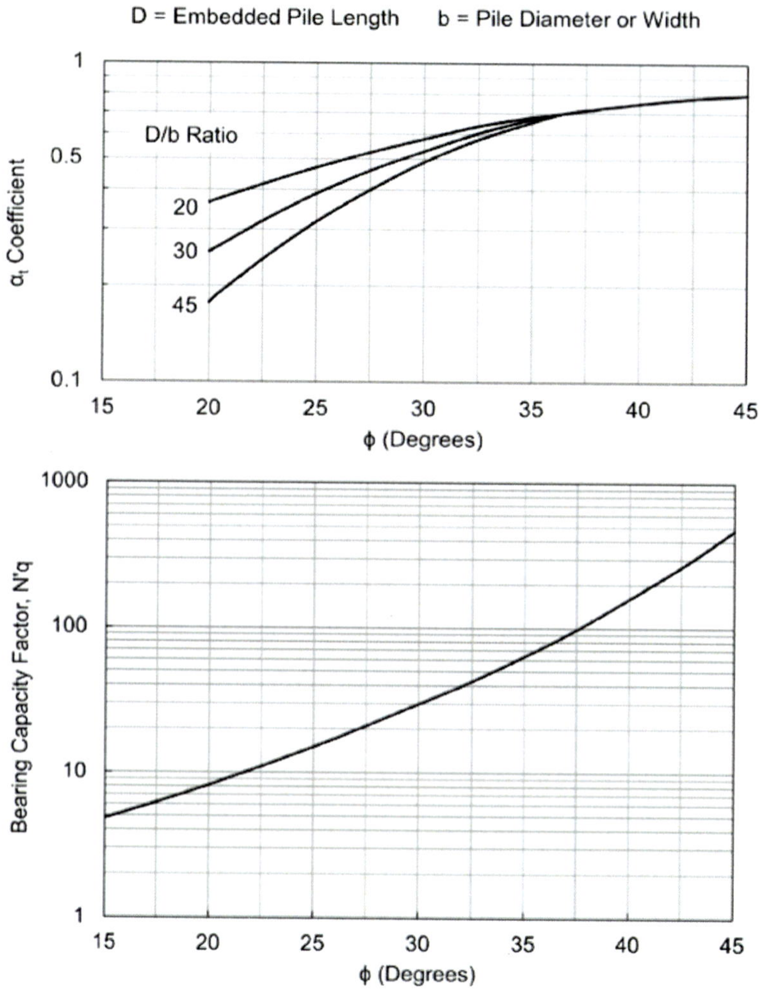

Fig. 7 Graphs for estimating α coefficient and N_q (FHWA GEC 012, 2016)

It can be clearly visualised from the combined charts in Fig. 1 that the values of N_q provided in IS 2911:1 Sects. 1 and 3 (2010a, c) is very close to the values given by Berezantsev et al. (1961) up to $\phi = 40^{\circ}$ and the remaining plot is close to that provided by Skempton (1951). On observing Fig. 2, it can be said that the values of N_q provided in IS 2911:1—Sects. 2 and 4 (2010b, d) are similar to those provided by Thurman (1964) and, hence, this could be the possible source for N_q factor in IS 2911:1—Sects. 2 and 4 (2010b, d). From Fig. 3, it can be seen that the values of α, in German Code DIN 4014 (1990) and in all sections of IS 2911: Part 1 (2010a, b, c, d) are similar.

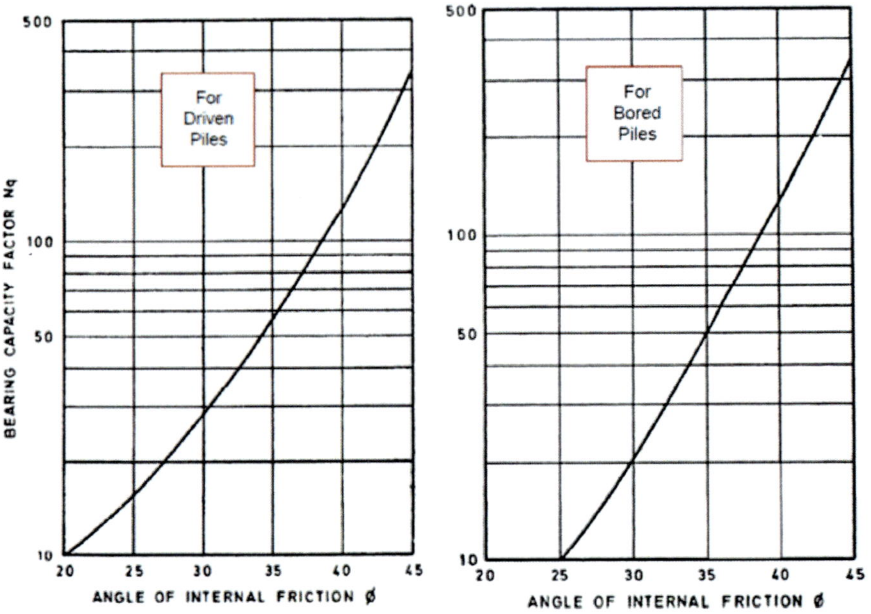

Fig. 8 Nq versus ϕ (IS 2911 Part 1: Sect.1–4, 2010a, b, c, d)

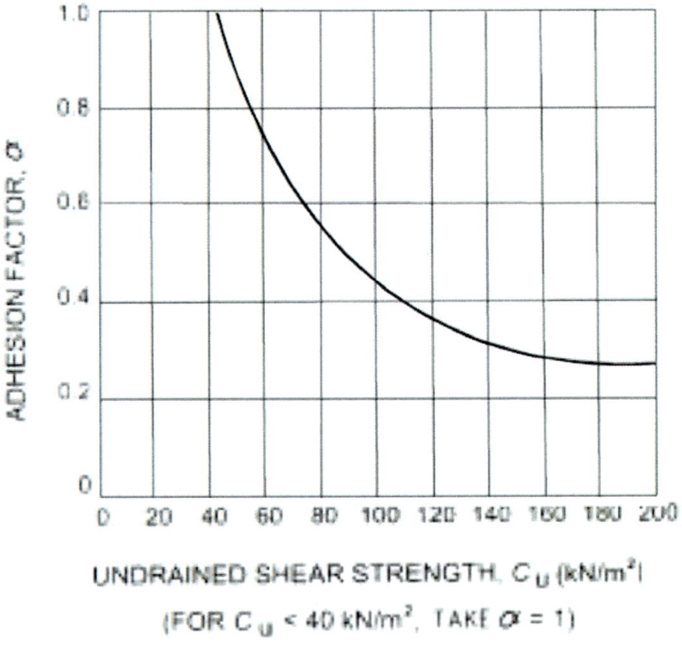

Fig. 9 Variation of α with C_u (IS 2911 Part 1: Sect.1–4, 2010a, b, c, d)

5 Regression Analysis

As IS 2911:1 (2010a) makes use of graphs for the computation of N_q and α, it may not be considered as an effective way in the associated studies demanding the automation of the procedure. The manual input of values could be cumbersome and erroneous especially for problems involving a large number of calculations. The designer has no scope for errors where important projects are involved and, therefore, to provide a solution, equations are developed in the present study using non-linear regression for computing the values of both the factors for a fast and reliable calculation. These equations are developed using NCSS software. Numerous non-linear models have been attempted and the ones with coefficient of determination greater than 0.99% are finalised. The finalised regression models are mentioned below in Eqs. 7–9.

(a) IS 2911 Part 1: Sects. 1 and 3: Driven Cast in Situ

$$\text{BearingCapacity Factor, } N_q = e^{\frac{0.824+0.0486\times\phi}{1-0.011\times\phi}} \tag{7}$$

where ϕ is in Degrees.

(b) IS 2911 Part 1: Sects. 2 and 4: Bored Cast in Situ

$$\text{BearingCapacity Factor, } N_q = e^{\frac{-0.628+0.099\times\phi}{1-0.0077\times\phi}} \tag{8}$$

where ϕ is in Degrees

(c) IS 2911 Part 1: Sects. 1–4

$$\text{Adhesion Factor, } \alpha = 66.83 - (42.53 \times \tan^{-1} C_u) \tag{9}$$

where $\tan^{-1} C_u$ is in radians and C_u is the undrained shear strength defined in kN/m^2. The maximum value of adhesion factor, $\alpha = 1$, for all practical purposes. Equation 9 is valid in the range for $C_u > 40$ kN/m^2 and $C_u < 200$ kN/m^2, respectively.

6 Conclusions

The comparison of methods of bearing capacity factor and adhesion factor done in this paper highlights possible sources from which the method given in the Indian Standard for Pile Design may have been adopted as guidelines for the design of pile foundations.

N_q, provided in IS 2911 1—Sects. 1 and 3 (2010a, c) is adopted from Berezantsev et al. (1961) up to $\phi = 40°$ and the remaining portion seems to be taken from Skempton (1951). Furthermore, N_q provided in IS 2911:1 Sects. 2 and 4 (2010b, d) is similar to that provided by Thurman (1964). DIN 4014 (1990) and IS 2911 (2010a, b, c, d) provisions for adhesion factor (α) are found to be slightly similar. Also, accurate and straightforward equations have been proposed to estimate N_q and α. The developed multivariable non-linear regression models can be used with ease to automate the analysis of pile foundations.

References

AASHTO LRFD (2017) Bridge design specifications, 8th edn, Section 10: Foundations

API RP 2A-WSD-2007 (2007) Recommended practice for planning, designing and constructing fixed offshore platforms-working stress design

BS 8004 (2015) Code of practice for foundations. BSI (Bristish Standards Institution), London

BS EN 1997: Part 1: (2004) Eurocode 7- Geotechnical Design Part 1, British Standard

Berezantsev VG, Khristoforov VS, Golubkov VN (1961) Load bearing capacity and deformation of piled foundations. In: Proceedings of the 5th international conference on soil mechanics and foundation engineering, vol II, Paris, pp 11–15

Bowles JE (1977) Foundation analysis and design, 2nd edn. McGraw-Hill Book Company, New York, USA

Caquot A, Kerisel J (1956) Traite de Mechanique des Soils, 2nd edn. Gauthier Villers, Paris

Deutsches Institut fur Normung (DIN 4014) (1990) German association for earthworks and foundation engineering

FEMA P-55/Volume I/(August 2011): Coastal construction manual

FHWA NHI-16-072 (2016) NHI Course No. 132031, Geotechnical site characterization

Hansen JB (1951) Simple statical computation of permissible pile loads. Christ Nielsen Post:14–15

Hansen JB (1961) A general formula for bearing capacity. Danish Geotech Instit, Bull 11:38–46

IS 2911 Part 1/Sect 1 (2010a) Design and construction of pile foundations part 1 concrete piles; section 1, Driven Cast In-Situ Concrete Piles

IS 2911 Part 1/Sect 2 (2010b) Design and construction of pile foundations part 1 concrete piles; section 2, Bored Cast In-Situ Concrete Piles

IS 2911 Part 1/Sect 3 (2010c) Design and construction of pile foundations part 1 concrete piles; section 3, driven precast concrete piles

IS 2911 Part 1/Sect 4 (2010d) Design and construction of pile foundations part 1 concrete piles; section 4, precast concrete piles in prebored holes

Janbu N (1976) Static bearing capacity of friction piles. In: Proceedings of the 6th european conference on soil mechanics and foundation engineering, vol 1–2

Kulhawy FH (1991) Drilled shaft foundations. In: Foundation engineering handbook (pp 537–552). Springer, Boston, MA

Meyerhof G (1953) The bearing capacity of foundations under eccentric and inclined loads. In: Proceedings of the 3rd international conference on SMFE (vol. 1, pp 440–445)

Nordlund RL (1963) Bearing capacity of piles in cohesionless soil. JSMFD ASCE 89(SM 3)

Poulos HG, Davis EH (1980) Pile foundation analysis and design

Skempton AW (1951) The bearing capacity of clays. Building Research Congress. London. The Inst. of Civil Eng., Div. I, 180p

Terzaghi K (1943) Theoretical soil mechanics. Wiley and Sons, New York, USA

Thurman AG (1964) Discussion of "bearing capacity of piles in cohesionless soils" by R.L. Nordlund. J Soil Mech Found Div, ASCE 90(1):127–129

Tomlinson MJ (1986) Pile design and construction practice, E & FN SPON, Palladian

Vesic AS (1963) Bearing capacity of deep foundation in sand. Highway Research Record No. 39, Highway Research Board, Washington DC, USA

Vesic AS (1977) Design of pile foundations. NCHRP Synthesis of Highway Practice (42)

How Choice of Foundation Can Alter the Fate of a Bridge Project—Three Case Studies

Alok Bhowmick

Abstract The most uncertain and challenging part of a bridge design and construction is the 'Choice of foundation'. During the planning and conceptual design stage, it is extremely important to choose the right type of foundation. Wrong choice can lead to disaster for the project. This paper will present three interesting case studies of past bridge projects, to demonstrate how correct choice of foundation can help to meet the project commitments, while the wrong choice of foundation can lead to a huge delay in project completion, leading to time and cost overrun.

Keywords Pile foundation · Well foundation · Prestressed rock anchors · Vibro-hammer · Jacketing

1 Introduction

The first case study will be for a project in Delhi, demonstrating the correct choice of foundation. The project is the twin bridge over the River Yamuna on either side of the existing Nizamuddin Bridge, for the Delhi–Meerut Expressway project. The design and construction of this bridge could be completed in a record time of 360 days. The key reason for this achievement was the correct choice of foundation.

The second case study will be another bridge project in Delhi over the same River Yamuna, where an inordinate delay in construction took place due to the wrong choice of foundation. This is the Railway Bridge over the River Yamuna at Delhi located at 30 m upstream of the existing rail-cum-road bridge. The foundation of this proposed bridge was initially planned with well foundations and sub-structures of eight piers and one abutment were completed with a well foundation in the year 2007. Wells at locations of five piers could not be sunk up to designated levels due to the presence of rocky strata, which led to the introduction of pile foundation as an afterthought. These foundations are still stuck for completion even after 13 years.

A. Bhowmick (✉)
Managing Director, B&S Engineering Consultants Pvt. Ltd., Noida, UP, India
e-mail: bsec.ab@gmail.com

B. Adimoolam et al. (eds.), *Deep Foundations for Infrastructure Development in India*, Lecture Notes in Civil Engineering 315, https://doi.org/10.1007/978-981-19-8598-0_8

The third project to demonstrate wrong choice of foundation is the Bagchhal Bridge in Himachal Pradesh, which is a balanced cantilever-type bridge connecting Swarghat and Marotan and is having a span configuration of 60 m + 185 m + 72.5 m. The bridge construction started in the year 2005 and is partially constructed. Work is stuck due to multiple problems at site; the primary reason being a faulty foundation.

2 Case Study 1: Twin Bridges Over the River Yamuna Adjacent to the Existing Nizamuddin Bridges, Delhi—A Success Story for Fast Construction

This case study deals with the planning, design and construction of the twin bridges which were constructed by NHAI adjacent to the old Nizamuddin bridges of PWD. These bridges were part of the Delhi–Meerut Expressway corridor. The bridges have several unique features, most notable amongst them is the fact that the design and construction of this bridge have been completed in a record time of 357 days. This project was awarded to concessionaire 'M/S Welspun Delhi Meerut Expressway Private Limited' at a total project cost of ₹841.50 crores. Of the total project cost, 40% is funded by the National Highways Authority of India (NHAI) under the new 'Hybrid Annuity Model' scheme.

The construction of this project commenced on 1 November 2016. Subsequent to award of work, the project got delayed due to the declaration of the River Yamuna as National Waterway by GoI under the National Waterways Act, 2016. In the light of this unprecedented development, the Yamuna Action Committee (or NGT) announced that any upcoming bridges over Yamuna downstream of the Wazirabad barrage shall provide for lateral and vertical navigable clearances as per the norms of Inland Waterway Authority (IWAI). This was a major technical and contractual variation in the project. In order to finalize the mode of change of scope, a significant working time was consumed, which influenced all the stakeholders involved in the project to further squeeze the construction time for the project. The design parameters including the vertical and horizontal clearances were frozen only in December 2016, nearly after a year, enabling the commencement of bridge design in January 2017. Actual commencement of work at the site commenced on 1 March 2017 and the project was completed in February 2018, in a record time of less than a year.

3 Secret of Success

It was clear from the very beginning that success of the project lies in reducing the construction time for the bridge. Right since the onset of the project, the executing agency had decided to adopt such a scheme, which allows the least time of construction without any uncertainty. With this objective, the following decisions were taken after brainstorming between the design consultant and the execution agency :

a. It was decided to adopt pile foundation (for piers) instead of traditional well foundation so that uncertainties (attached to sinking of wells) and time-intensive work of well foundation is minimized. This was one of the most crucial decisions taken early in the project, which helped completing the project in a record time. It is worth noting that the change of foundation system from 'well' to 'pile' had increased the initial cost of the bridge by approx. 8 crores, though it helped significantly in reducing the construction time and thereby the overall cost of the project.
b. Very strategically, it was planned to complete all piling activities by 3 July 2017 and demobilize all machinery and equipment from an active course of river. The planning and management of execution were done to near perfection to ensure that the last piling was cast just about 12 h before a huge quantum of discharge was released from upstream barrage.
c. For superstructure, large-scale use of precast girders is envisaged so that casting of girder can continue in all seasons as a parallel activity to casting of sub-structure and foundation. The possibility of using a launching truss for erection of precast girders was explored. But this was ruled out since use of launching truss would force erection on a defined sequence from one end to the other end and, in the worst scenario, if any one foundation was stuck due to any unforeseen reasons, the entire chain of activities will be affected. Since the river stream in lean season is very much restricted and crane movement for almost the entire bed of Yamuna is feasible for 9 out of 12 months of the season, it was decided to erect the girders by using double cranes, so that, wherever any two adjacent piers (of any carriageway) are ready, girder erection can commence. Heavy crawler-mounted cranes of up to 450t capacity were mobilized for this task.
d. Both, precast post-tensioned as well as pre-tensioned girders have been used successfully. This combination of pre- and post-tensioned girders was necessitated due to the non-availability of adequate space for casting yard for pre-tensioned girders. Initially, all girders were envisaged as pre-tensioned girders. It was decided to go for post-tensioned girders for the LHS bridge and pre-tensioned girders for the RHS bridge. Post-tensioned girders were cast in river bed and transported to the location using a special multi-axle low-bedded trailer.
e. Due to the possibility of deep scour in the river, it is mandatory to provide steel liner up to the scour depth. To speed up the driving of MS liner, vibro-hammer was deployed at site. Further, in order to facilitate ease of driving of the liner by 8MT capacity vibro-hammer, the thickness of liner was increased from 6 mm (as per the minimum requirement of code) to 8 mm.
f. It was realized quite early in the project that bearings and expansion joints have a long delivery time due to high demand and restricted number of quality suppliers.

The bearing layout and forces were therefore finalized early in the design development stage and the order was placed to the vendor much in advance. The delivery of bearings could not delay the erection of girders.

4 Span Arrangement and Structural Scheme Finally Adopted

Dictated by the existing bridge span arrangement and length, the span arrangement for the bridge comprised 13 spans of 42.4 m each. Expansion joints are provided at alternate piers for end modules and, at every third pier for intermediate modules. Figure 1 shows the General Arrangement of the Bridge. The bridge is having dual carriageway of 15 m each (four lanes) with cycle track and footpath. The two carriageways are placed on either side of the existing pair of bridges at the location. The overall width of the bridge is 20.5 m each [0.5 m (Railing) + 1.5 m (Footpath) + 2.5 m (Cycle Track) + 0.5 m (Crash barrier) + 15 m (C'way) + 0.5 m (Crash Barrier)]. Figure 2 shows the typical cross section of the bridge. The superstructure is a composite section comprising a cast-in-situ reinforced concrete deck slab, resting over precast PSC girders. The deck slab is made continuous over intermediate piers, thus avoiding expansion joints at every pier. The depth of PSC T Girder has been kept as 2.65 m. Self-weight of each post-tensioned girder has been restricted to 130 tons from consideration of ease of erection and handling. In case of pre-tensioned girder, the self-weight is lesser. Each four-lane carriageway of superstructure is supported on a cantilever RCC circular-type pier of diameter 2.7 m. The rectangular pier cap of dimension 3.4 m × 18.75 m is provided over the circular pier. The pier is resting on bored cast-in-situ pile foundation. Six numbers of 1.8 m diameter bored cast-in-situ piles under each pier have been provided. These piles have been taken down to 45 m below the cut-off level. The safe load carrying capacity of pile considered is 645 tonnes. The pile cap is 2.7 m thick of dimension 7.6 m (Long) × 13.0 m (Transverse).

5 Construction Speed

All construction activities were monitored on a daily basis to ensure quality as well as speed of construction. The entire piling work was completed in 18 weeks, with an average speed of 8 piles per week. Construction of well foundation took about 36 weeks, but since this activity was not on the critical path, this did not affect the overall progress. Girder casting was accomplished in 30 weeks with an average speed of about six girders per week. Deck slab concreting for 26 spans was accomplished in a record time of 20 weeks. Table 1 gives the progress chart of different construction activities in the project.

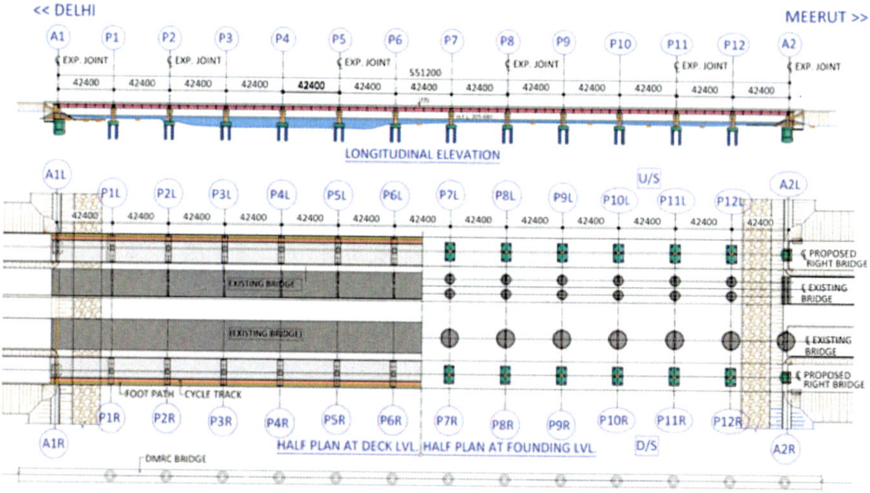

Fig. 1 General arrangement of the bridge

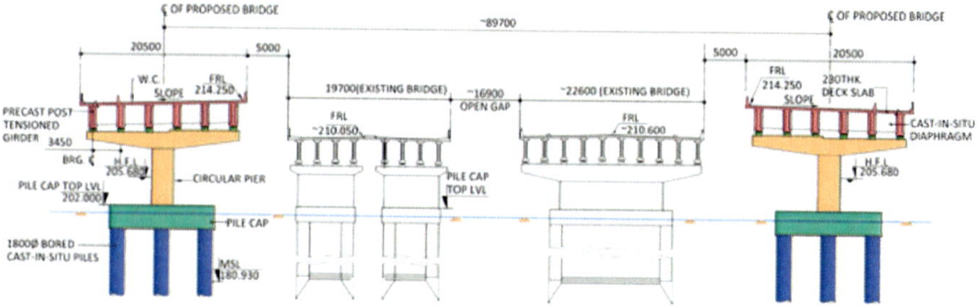

Fig. 2 Typical cross section of the bridge

6 Key to Success and Lessons Learnt

Initial conceptualization of the structural scheme and planning for execution is extremely important in the success or failure of a project. It is also extremely important to continuously monitor the progress and, in case of any deviation from the original plans at any stage due to any unforeseen reasons, prompt action is needed in modifying the original plan. Some of the key reasons for the success of this project are as follows:

a. Advanced resource planning and continuous monitoring by the Independent Engineer as well as NHAI officials.
b. Mobilization of extra manpower as well as equipment resources as and when needed, without any compromise on the speed of work.
c. Use of pile foundation for pier supports, despite the fact that the initial cost of pile foundation is more as compared to well foundation.

Table 1 Construction work progress chart

Task No	Name of Work	No. of Week	TIME IN WEEKS (STARTED FROM 01.03.2017 TO 28.02.2018												
			1 - 4	5 - 8	9 - 12	13 - 16	17 - 20	21 - 24	25 - 28	29 - 32	33 - 36	37 - 40	41 - 44	45 - 48	49 - 52
1	PILING WORK: (01.03.2017 to 30.06.2017) (144 Nos., 1.8m Dia, 45m Length)	18						8 Piles per Week							
2	WELL FOUNDATION: (08.03.2017 to 07.11.2017) (4 Nos., 8m Dia)	36													
3	GIRDER CASTING: (04.06.2017 to 30.12.2017) (156 Nos., 30 Pre-tensioned & 126 Post-tensioned)	30													6 Girders per Week
4	PILE CAP CASTING: (28.04.2017 to 03.01.2018) (24 Nos., 13mx7.6mx2.7m)	36													
5	PIER CAP CASTING: (25.06.2017 to 12.01.2018) (24 Nos.)	29													
6	LAUNCHING OF GIRDER: (30.08.2017 to 22.01.2018) (156 Nos.)	22													
7	DECK SLAB CONCRETING: (29.09.2017 to 10.02.2018) (26 Spans)	20													
8	EXP. JOINT INSTALLATION: (01.12.2017 to 15.02.2018) (12 Nos., 246m Length)	12													
9	CRASH BARRIER & WEARING COAT: (21.11.2017 to 17.02.2018)	14													

Photo 1 Bird's eye view of bridge during construction

Photo 2 Precast girder erection in progress

d. Prompt issuance of GFC drawings by the design consultant for execution at site as per the project demands.
e. Tailor-made design to suit the project-specific requirements.
f. Timely payment to all vendors by the executing agency.
g. Working round the clock (24X7).
h. Motivation from client during the execution.

Photos 1 and 2 shows the view of the bridge during construction.

7 Case Study 2: Construction of New Railway Bridge Over River Yamuna Adjacent to Old Rail-cum-Road Bridge, Delhi—A Case of Wrong Choice of Foundation Leading to Inordinate Delay

The Northern Railway is constructing a new bridge over the Yamuna River in Delhi at 30 m upstream of the existing rail-cum-road bridge No. 249 on Delhi–Delhi Shahdara section of the Delhi Division for double railway line and 25 MT loading. The construction began in 2003. Seventeen years have passed and the bridge is still not completed. As of now, only 50% of the work seems to be completed. The latest deadline for completion of the work is December 2021.

The existing rail-cum-road bridge was built in 1867 at a cost of Rs. 16 lacs and has served humanity well for over 150 years (Photo 3). Initially, the bridge had a single railway line, and was converted to a double line in 1913. The bridge was designed for the highest flood level of 204.8 m, which is much lower than the current maximum design flood level of 208.5 m. The existing bridge had to be closed to traffic for the first time in 1956 when the water level touched 206.4 m level. It was closed again in 1978 when the flood level reached 207.6 m. Since 1995, the bridge is frequently closed during monsoon season as a precautionary measure whenever the flood level reaches the danger level. Also, trains are allowed to pass over this bridge only at a restricted speed of 30 Km/hr. A new bridge to replace the existing bridge is therefore in demand for quite a few years now. Well foundation of this existing bridge is made of brick masonry. Superstructure is of open web steel girders with a span arrangement comprising 12 spans of 61.0 m (200') each. The new bridge in replacement was sanctioned in 1997–98 at an abstract cost of Rs. 58.32 crores.

Originally, 13 spans of 61m each were envisaged. But due to the subsequent decision to realign the bridge, the span arrangement was revised, and finally, it was decided to have the new bridge with 14 numbers of spans with span arrangement of 11 x 61.0 m Open web steel girders + 2 x 45.70 m Open web steel girders with deck slab + 1 x 30.50 m Composite steel girder. The sanctioned cost was revised in 2011–12 from Rs. 58.32 crores to Rs. 136.95 Crores.

The foundation of this proposed bridge was initially planned with well foundations. Construction started in 2003 and sub-structures of eight piers and one abutment (A-1, P1–P7) were completed with well foundation in the year 2007. Wells at five locations of piers P8–P12 could not be sunk up to designated levels in rocky strata. These partly constructed well foundation also had undergone excessive tilt and shift due to sloped rock at the well cutting edge level. Detailed scrutiny by experts revealed that the bedrock surface in this area is having a slope and its level varies from pier to pier. The rock is mainly of crystalline medium- to coarse-grained Quartzite. At some locations, medium- to fine-grained Mica Schist has also been encountered. The

Photo 3 Existing 150 year old rail cum road bridge

rocky strata is sloping down towards P7 both from A1 and A2. Alignment of bridge is passing through ridge, which is part of the Aravali range of mountains and not on isolated rock mass.

Meanwhile, the alignment of the bridge approach towards Delhi side came into a dispute with the Archaeological Survey of India (ASI) since as per the planning of the Northern Railway, a new rail track was to put loads on the sixteenth-century world heritage site of Salimgarh Fort. In 2012, decision was taken to realign the rail track and bridge so that the alignment will merge with the existing track over the fort, in order to protect the national monument. In the process, the well P12 became defunct. Revised span arrangement in this zone had to be framed. Two spans of 47.85m each and a single span of 32.45 m replaced the previous two spans of 61m. New supports, P12 and P13 created which are proposed to be supported on pile foundation.

After much deliberations, the Northern Railway decided to abandon the partly constructed and excessively tilted/shifted well foundations at these locations and provide large diameter piles inside these wells to support the structure. Accordingly, 1500 mm dia piles were constructed at location of pier P8–P11. Due to the provision of pile foundation at P8–P11, the wells have become redundant and are not being used for load bearing purpose. As piles are being done inside these redundant wells, it became necessary to secure these tilted and shifted wells in their position so that there is no further movement during service life of bridge to ensure safety of piles. Various possible options for securing the partly constructed well foundation were explored. Some of these are given below:

a. Providing prestressed anchors through holes in steining;
b. Providing high-strength metal bars through holes in steining;
c. Providing external piles;
d. Dismantling wells; and
e. A combination of the above methods.

None of these methods were finally chosen and, after much deliberations, it was realized that for securing the well foundations, the solution cannot be same in all wells since the ground situation in every well foundation is different. Well-specific solution was evolved. For well foundation at P8 and P11, the maximum gap between well kerb and bedrock was found to be about 1.60m. These two wells are proposed to be secured by providing a bottom plug up to 150 mm above well kerb. Piles are proposed to be isolated from concrete of bottom plug by providing MS sleeves keeping a minimum 100 mm gap all around piles.

For P9 and P10, the maximum gap between well kerb and bedrock is nearly 5m and 6m, respectively. These two wells are therefore proposed to be secured by providing bottom plug with M-25 grade concrete up to 150 mm above well kerb. Piles are proposed to be isolated from concrete of bottom plug by providing MS sleeves keeping a minimum 100 mm gap all around piles. After bottom plugging, stabilization of sand below bottom plug is proposed to be carried out by grouting suitable material through slotted pipes. These slotted pipes are to be provided in concrete (protruding as deep as possible) before casting of bottom plug. Photos 4 and 5 show the aerial view of construction in progress at site.

Photo 4 Abandoned well foundation of new bridge and existing bridge

Photo 5 Drone view of new under construction bridge & old Bridge

Lessons learnt:

Some of the key reasons for this inordinate delay in the construction of this bridge are as follows:

i. Poor quality and extent of geotechnical investigation: Initial geotechnical investigations carried out by client were not appropriate—both quantitatively as well as qualitatively, which lead to the lack of proper understanding about the sub-strata.
j. Wrong choice of deep foundation: Large diameter pile foundation should have been the original choice in this bridge, considering the fact that rock was encountered at variable depths, at the time of initial investigations.
k. Delayed decisions during construction: When problem was encountered in sinking of well foundation initially by the contractor, some of the decisions appear to have been taken in haste, causing further delay in the construction process. As an example, the decision to construct pile foundation inside the dredge hole of a partly constructed well was taken in haste without really thinking about how to

safeguard and protect the well foundation once the pile foundation is constructed. This led to further delay.

Important take away from this project is:

A thorough initial investigation, making an appropriate choice of foundation by senior bridge experts at the inception stage and taking collective decisions during the construction stage in case of any crisis are the prerequisites for the success of any project.

8 Case Study 3: Construction of Bridge at Bagchhal Over Govind Sagar Reservoir Near Swarghat, Bilaspur (H.P)—A Case of Wrong Choice of Site, Structure Type and Foundation Type Leading to Time and Cost Overrun

Background

The administrative approval and expenditure sanction for this bridge was accorded for Rs. 1891.10 Lacs more than 25 years ago, way back on 11 May 1994. The work was awarded to a contractor on a lump-sum tender basis, more than 20 years ago, in the year 2005 for a cost of Rs. 2283.52 lacs and the scheduled date of completion for this bridge was 4 July 2008. The scope of work involved the construction of 317.50m span superstructure, 3 spans continuous with span arrangement of 60m + 185m + 72.5m, having two deck monolithic with piers and anchored at two abutments.

Figure 3 shows the General Arrangement drawing of the bridge which was originally envisaged. Figure 4 shows the cross section of the bridge. The tender stage drawing for the bridge envisaged a founding level of piers P1 and P2 at R.L 460.0m.

The construction activity started in full blast in 2005. On the basis of geotechnical investigations carried out by the contractor during construction, the founding level

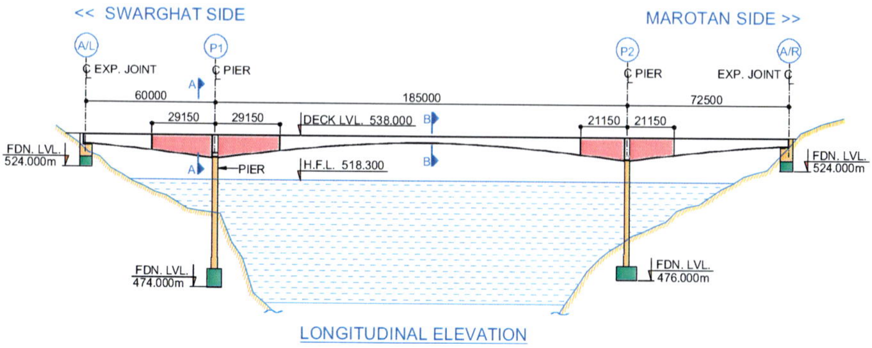

Fig. 3 General arrangement of the bridge as per original scheme

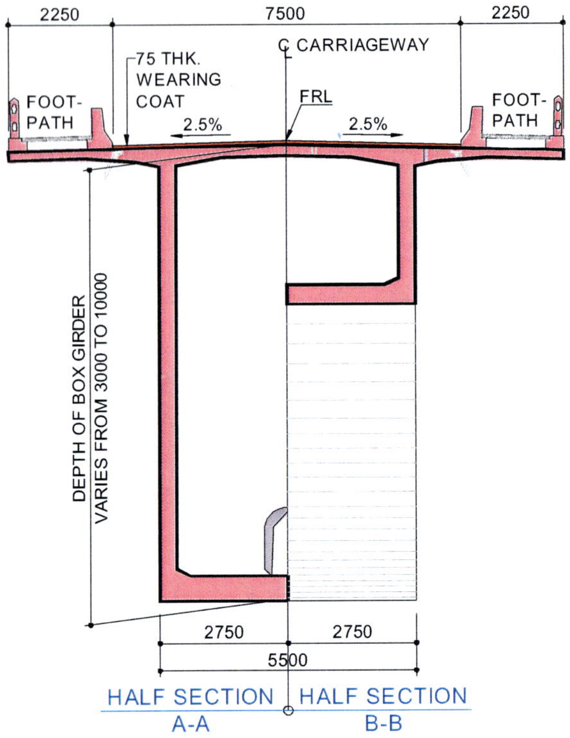

Fig. 4 Cross section of superstructure

for the piers was raised from R.L 460.0m (as given in the tender stage drawings) since this level was considered too deep looking at the actual sub-strata. After deliberations between client and Contractor, the founding level was fixed as R.L 474.0m for P1 and R.L 476.0m for P2.

The construction of foundations at pier P1, P2 is completed with the raised level and also the piers over them are completed. Even a part of the superstructure was constructed using the cantilever construction technique. Photo 6 shows the partly constructed bridge. Meanwhile, in view of the unforeseen ground conditions and fragile nature of the geology of the area, advice of geotechnical and geological experts was sought by the executing agency to study the suitability of the foundation as well as the stability of the hill slopes near the bridge site. The report from the geological and geotechnical experts indicated an extremely heterogeneous nature of rock formation in the area and also revealed many unpleasant surprises. It was discovered that there is a fault plane which crosses the bridge alignment, which thereby raised the question of suitability of the alignment itself. Considering all these uncertainties and doubts regarding stability of the already constructed foundations at P1 and P2, the construction was stopped in January 2010.

Photo 6 Portion of structure completed in 2010, before the work was suspended

Revival of the Project in December 2016

After 11 years, the work on this bridge starts again. In 2016, the possibility of salvaging of the bridge and its revival was brain-stormed between the executing agency and client. Opinion of Dr V K Raina, a renowned bridge expert and a World Bank Consultant, was taken in this regard at this point. It was decided to restart construction work on this project, strengthen the foundations, review the design of sub-structure and foundation by an independent agency and complete the project. Work restarted in December 2016 and the following major modifications were identified:

a. **Modification in Span Arrangement:** The structural arrangement is slightly modified at Abutment A/L. Since at the abutment location, the vertical pre-stressing steel installed in the initial construction phase was found to be corroded, the existing abutment, designed to take uplift is converted into a pier, which will be designed to take downward load rather than uplift. The 60 m span of superstructure is extended beyond A/L by additional 12.5 m and new abutment A0 is proposed.
b. **Salvaging of Foundations at Piers P1 & P2:** For Pier P1, prestressed rock anchor is added to the existing foundation. For P2, the existing shaft foundation is being strengthened and piles are constructed to transfer the load to deeper strata.
c. **Jacketing of Piers:** Concrete jacketing is proposed around the piers to strengthen the piers.
d. **Modification in Superstructure Design:** Minor modifications are carried out in SIDL to reduce the load from top. A combination of external prestressing and internally bonded tendons have been used in the final approved design.

Figure 5 shows the revised general arrangement drawing for the project. Photo 7 shows dowelling of the existing pier for jacketing. Photo 8 shows piling at P2 and Photo 9 shows the work on pile cap at P2.

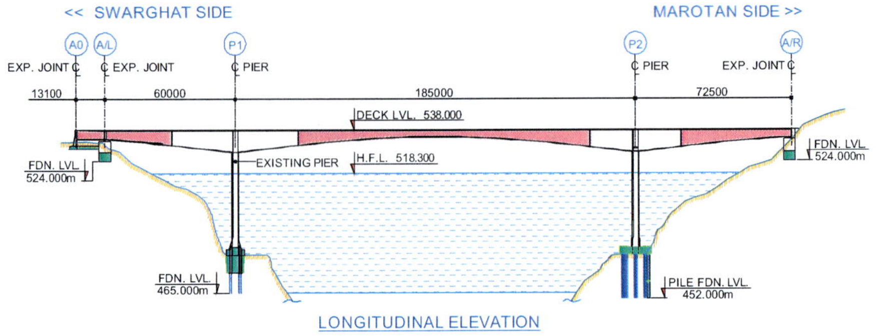

Fig. 5 General arrangement of the bridge as per revised scheme

Photo 7 Dowelling in pier P1 in progress for jacketing of Pier

At P2, it is proposed to have ten numbers of 1.2m diameter piles which are socketed 30m deep from top of shaft into rock with a pile tip level of R.L 452.0m, as per the decision taken at the time of agreement. Construction of bored pile socketed so deep into rock could be possible only by using a combination of specialized piling equipment like MAIT and RCD (Reverse Circulation Drilling) system, also known as air-lift system. Bringing such equipment to the project site itself was a challenge. A special temporary road had to be constructed to bring these equipment's safely to project site. The initial target was to complete the salvaging works of foundation and

Photo 8 Pile boring in progress at P2

Photo 9 Casting of pile cap in progress at P2

remaining works of substructure and superstructure by 2019 (i.e. in 36 months), but due to the unpredictability of water level at the location, work on foundation could not be progressed at the speed desired.

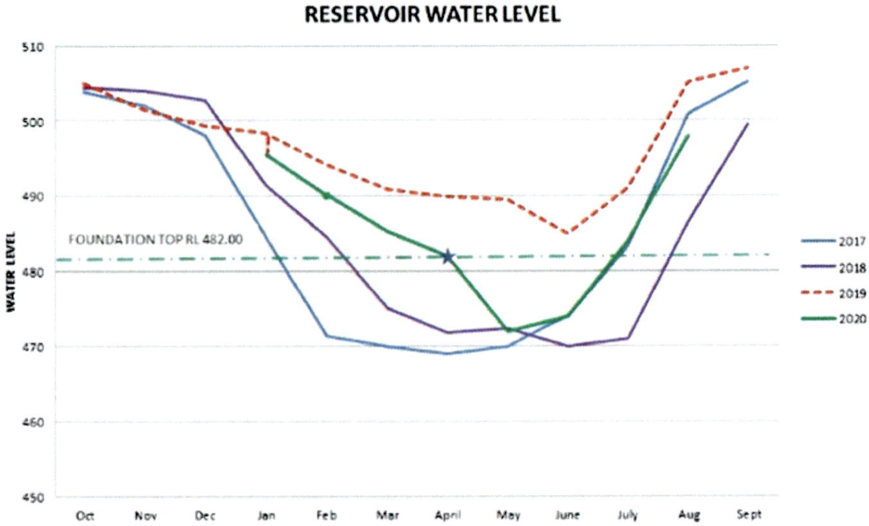

Fig. 6 Variation of water table at project site

Challenges in the construction of Pile Foundation at P2

It can be seen from the graph in Fig. 6 that the water level never came below the foundation top level in the year 2019. So, no work could be taken up at P2 for 22 months, from July 2018 to May 2020, when water level receded below the top of foundation for a brief period of 3 months only. The pile cap work is still incomplete and it would require one more season to bring the work above the minimum water level. Despite all the challenges—geological surprises, technical challenges, bureaucratic and administrative hassles—the project seems to be moving forward and one can see light at the end of the tunnel now. The project is targeted to be completed in December 2022, 17 years after it was commenced.

Lessons Learnt and the Way Forward:

Several key lessons can be learnt from the delayed execution of this iconic project. It highlights the need to carry out the geotechnical and geological investigation to the fullest at the inception of the project before even the project is tendered for execution. Right choice of foundation is a prerequisite for timely completion of such projects.

9 Conclusions

(a) Correct choice of foundation type for a given site condition, keeping in mind the available time of construction, is the most important decision in a project, which can decide the fate of the project. The three case studies presented above prove this fact.

(b) Need for proper geotechnical investigation at the feasibility stage of the project which can help in taking correct decisions on foundation type can not be overemphasized.
(c) It is very necessary to analyse the comparative merits and demerits, construction time frame and cost of construction of bridges with various possible types of foundations before finally choosing the type of foundation.

Soil–Structure Interaction for a Tension Pile Pulled with Strand

P. V. Chandramohan

Abstract A unique tension pile where the pull is at a point a certain distance below the top of the pile. The pull is transferred by a cable consisting of high-tensile strands. The HDPE duct of the cable prevents bond between the strands and pile concrete in the upper portion. While soil friction on the pile will act downwards on the whole length, there is tension in the bottom portion of pile and there will be compression at the top portion.

Keywords Tension · Compression · Annular space · High-tensile strands · Side friction

1 General

Tension piles are subjected to pull-out forces. The pull-out force will be transferred to the pile from the superstructure direct or from the pile cap. Tension is normally transferred through steel rods to the frame. Usually, tension piles are reinforced to cater to the tensile force. This is done by taking the reinforcement rods into the pile cap or superstructure directly. Tensile force in the pile is counteracted by the side friction between the soil and the pile. Side friction depends on the lateral soil pressure acting from all on the pile surface. Lateral pressure will go on increasing towards the bottom. This is given by the following expression:

$$P_D = K\gamma h \tag{1}$$

where

K—coefficient of earth pressure

γ—effective unit weight of the soil

h—depth below ground level.

P. V. Chandramohan (✉)
Navayuga Engineering Company Ltd., Hyderabad, Telengana, India
e-mail: pvcmohan@gmail.com

B. Adimoolam et al. (eds.), *Deep Foundations for Infrastructure Development in India*, Lecture Notes in Civil Engineering 315, https://doi.org/10.1007/978-981-19-8598-0_9

Please see Fig. 1a.

Side friction on the pile is the product of this pressure and the friction coefficient. This coefficient is taken as $tan\delta$, where δ is the angle of wall friction between the pile and the soil. This is illustrated in Fig. 1b.

Figure 1c shows the tension in pile when the pull on the pile and resistance from the soil is balanced. This is very simple. Stress in the pile is the tensile force divided by the cross-sectional area. This is given in Fig. 1d. There is also an intermediate case. When the applied load is less than the resistance of the soil, resistance of the full depth of soil is not required to resist the applied force. The extra length of the pile will be redundant. This is illustrated in Fig. 1e.

2 Unique Pile

The above are all simple usual cases, where the pull is applied at the top of the pile. The stress in concrete is maximum at the top. Towards the bottom, the downward friction of soil will go on reducing the pull-out force and so also stress in the pile. But there is a different type of pile where the pull is transferred not at the top of the pile. Please see Fig. 2.

At 1000 mm of the pile, there is an annular space. Purpose of the annular space is for the pile to accommodate horizontal movement of the load. The tensile force is applied to the pile through a cable consisting of high-tensile strands. These strands are flared out from the cable and embedded into the pile concrete. The strands are tied to the reinforcement cage of the pile before lowering the cage into the pile bore. The pile is 600 mm in diameter. The duct of the cable is made up of high-density polyethylene. Please see that the HDPE duct extends 500 mm downwards below the annular space. This 500 mm length is to act as a buffer against the progress of corrosion. The cable is taken to the top of the bottom slab of the superstructure and anchored there. The pile is to cater to a tensile load. It can be seen from Fig. 2 that the pull-out force will be transferred to the pile from the tip of the HDPE duct. This is almost 1500 mm from the actual top of the pile. Above this level, HDPE duct will prevent any bond developing between the pile concrete and the cable. Friction of soil on the pile will act on the entire length of the pile. Under the action of the pull, the whole pile will tend to move up. Side friction on the periphery will act in the opposite direction—downwards.

3 Transfer of Force

A schematic force/pressure diagram of the pile is given in Fig. 3a and b. The pull is applied at the bottom level of the HDPE duct. The portion of the pile below this level will be in tension as the side friction on the pile will act opposite to the pull. At the same time, the portion above this level will be under compression. Being, an integral

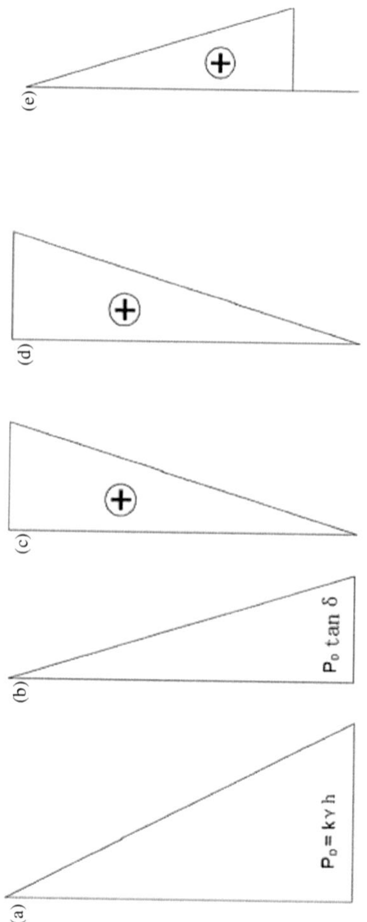

Fig. 1 **a** Lateral pressure. **b** Side friction. **c** Tension in pile. **d** Tensile stress. **e** Applied load < soil resistance

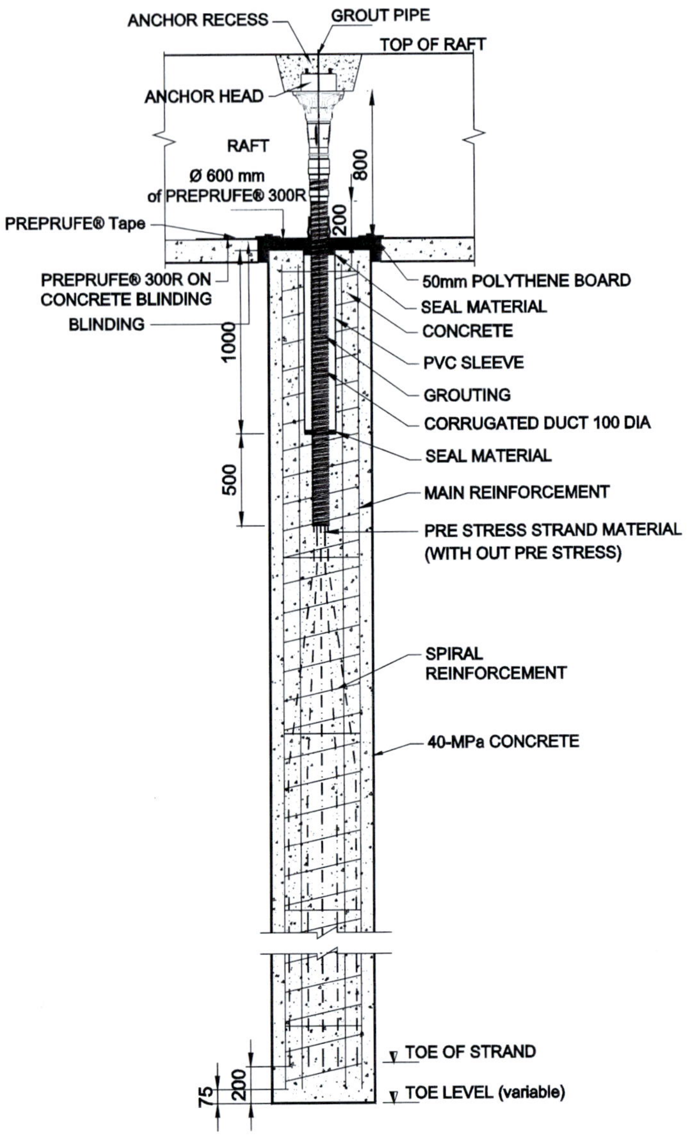

Fig. 2 Arrangement of strands and cable

part of the pile, the portion above also will move up. In other words, this portion will be pushed up by the portion below it. This will be resisted by the side friction which is downwards. Since the pushing force is at the bottom, this portion will be under compression. This pile is a part of a group of piles to support an underground reservoir in Qatar. Uplift on the piles comes from the uplift on the empty reservoir. A number of piles were load tested. Interesting data was obtained as the pile was instrumented. Levels of strain gauges are shown in Fig. 4.

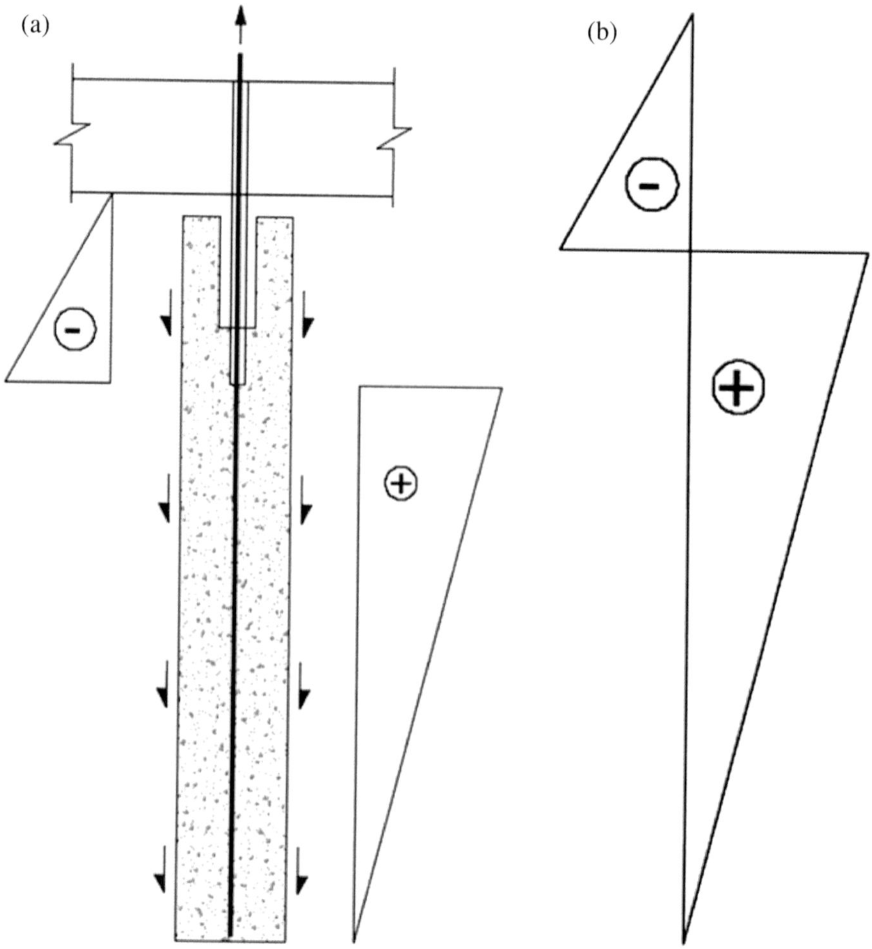

Fig. 3 **a** Schematic force diagram. **b** Combined force diagram

4 Strain Gauges

Please see that strain gauges were fixed at six levels in the pile from −1.5 to −5.8. The strain gauges were monitored at various stages of the load test. Four numbers displacement transducers were installed between a reference beam and the top of the pile at 90 degrees apart to monitor the displacement of the pile top.

Fig. 4 Location of the strain gauges

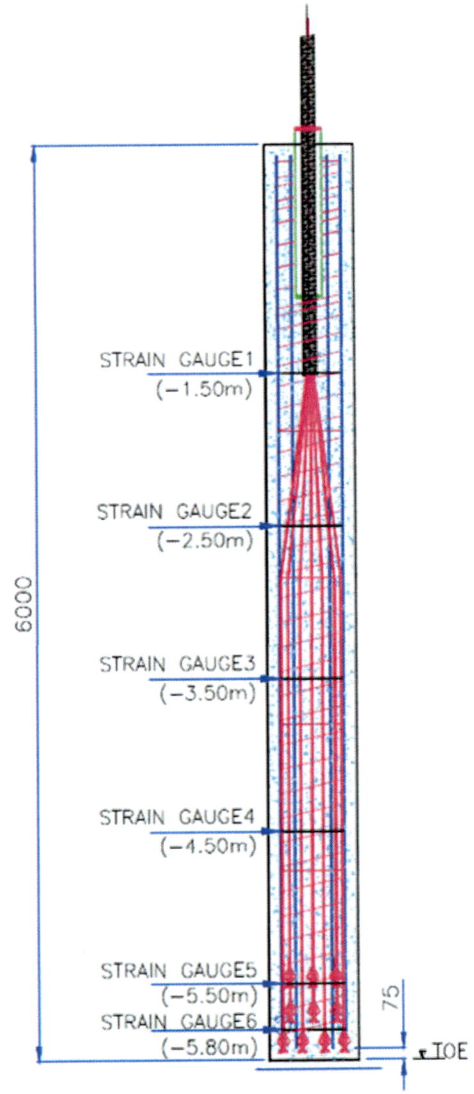

5 Geotechnical Data (Arab Centre and for Engineering Studies, 2015)

The subsoil consists of Simsima limestone, about 18 m thick. There are layers of Midra shale and Rus formation underneath this. Standard penetration tool could not penetrate into Simsima limestone and could be termed as 'refusal'. The rock had an RQD range of 0–95, with an average of 50. The underlying layer of Midra Shale had similar RQD values. The underlying layers of Fahahil Velates and Rus formation also had the same properties. Safe bearing pressure for spread foundation on rock arrived at 450 kPa.

6 Structural Data

Structural data of the pile are given below:

Diameter of the pile	= 600 mm		
Concrete strength (C40)	= 40 MPa	E_c	= 32400 MPa
Steel (E_s)	= 210000 MPa		
Area of cross section of pile	= 282600 mm^2		
Area of steel	= 4396 mm^2		

$$\text{Force at a cross section} = \text{Strain} * ((\text{Area of concrete} * E_c) + (\text{Area of steel} * E_s)) \quad (2)$$

The latter part of the equation works out to 1.0079E + 10 N.

Microstrains are obtained from the strain gauges and are tabulated below (Table 1).

Microstrain readings are plotted along the axis of the pile. The graph is given in Fig. 4 below.

The above readings were converted into loads using Eq. (1) and are given in Table 2 below.

Computed values of force in a cross section are given in Table 2.

Figure 3a shows the theoretical stresses in the pile. The theoretical profile shows a sudden reversal of stresses at level −1.5. The stresses are compressive above and tensile below this level. Stresses in Fig. 7 have been computed from strain gauge

Table 1 Strain gauge readings

	Microstrain at different levels*10E + 9				
Level	750 kN	1500 kN	2250 kN	3000 kN	3750 kN
	50%	100%	150%	200%	250%
0	0.00	0.00	0.00	0.00	0.00
−1.5	−7.27	−23.67	−33.85	−49.44	−67.2
−2.5	14.94	36.69	69.19	103.02	113.4
−3.5	3.32	7.00	9.71	11.71	17.3
−4.5	2.86	5.96	8.06	8.57	12.1
−5.5	0.81	1.57	1.82	2.21	3.0
−5.8	1.15	2.14	3.11	3.70	−4.9

Table 2 Load calculations

	Load calculation (kN)				
Level	750 kN	1500 kN	2250 kN	3000 kN	3750 kN
0	0.00	0.00	0.00	0.00	0.00
−1.5	−73.326	−238.611	−341.234	−498.306	−677.136
−2.5	150.593	369.336	697.359	1038.372	1143.197
−3.5	33.510	70.519	97.840	118.074	174.372
−4.5	28.833	60.120	81.289	86.428	122.249
−5.2	8.159	15.785	18.336	22.232	30.6047
−5.5	11.641	21.550	31.305	37.296	−49.58078

Fig. 5 Sample collected from Simsima drill

readings. They are discrete points along the axis of the pile. For example, there is one strain gauge at −1.5. The next one is at −2.5. The strain gauge at −1.5 will read the maximum compression even if maximum tension also occurs there. The strain gauge at −2.5 is certainly in the compression zone. Values in between are not reflected in the graph correctly. This is understandable. It was found that the movement of the top was 0.06 mm for the design load and 0.68 mm, 1.1 mm and 1.42 mm for 150%, 200% and 250% of the design load, respectively. Residual movement was found to be 0.22 m. This was against a structural requirement of 3.29 mm.

7 Conclusion

This is a unique case of a tension pile where the pull is transferred at a finite distance below the pile top. The pull is affected by flexible high-tensile strands. There is an annular space of 1 metre in length at the top through which the cable passes. This is to facilitate horizontal movement of the load. The HDPE duct is present for another 500 mm below the annular space cutting off any kind of bond between the strands

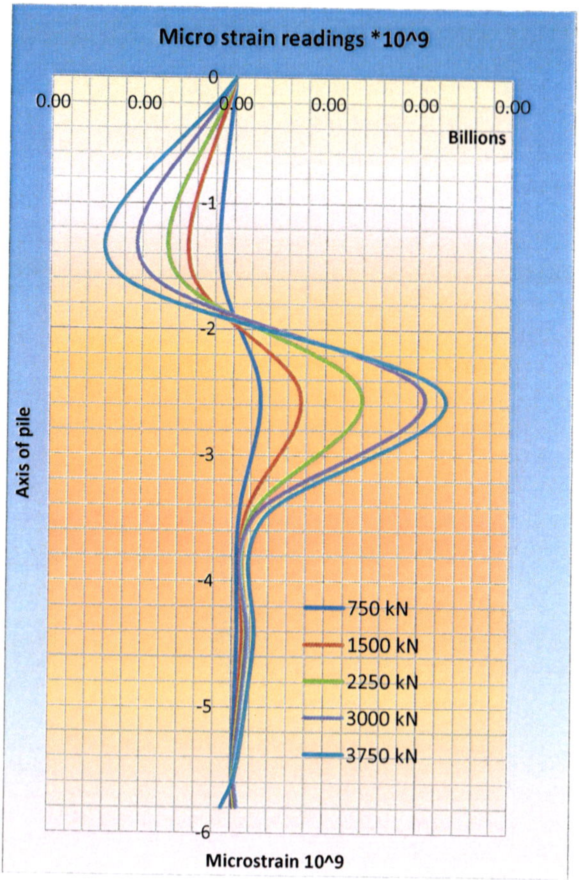

Fig. 6 Microstrain readings along the pile

and pile concrete. The pile was load tested to two and a half times the design load. The strain gauges installed in the pile provided an interesting picture of tension and compression along the pile axis.

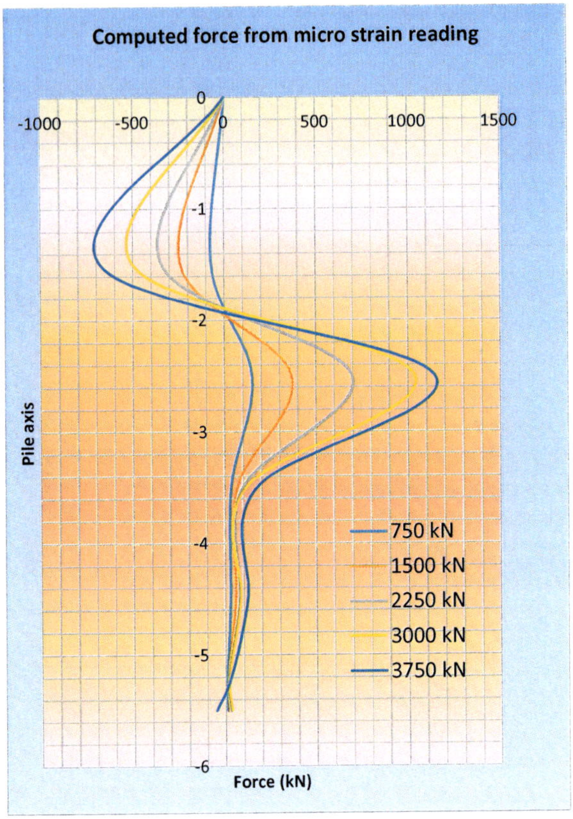

Fig. 7 Force along pile axis

Reference

Arab Centre for Engineering Studies (2015) Geotechnical site investigation for proposed GTC/518/2012 Kahramaa Al Muntazah, Qatar

Earth Retention and Deep Excavation Support

Challenges of Earthen Cofferdam in Deep Excavations for Waterfront Structures, a Case Study

P. V. Ramana, M. Balasubramani, and K. Bairagi

Abstract This paper is a case study of an earthen cofferdam, constructed for a deep excavation at intake location in River Aaundha, Maharashtra, India. A 12 m (39 ft) deep excavation was made to construct the intake well. The natural soil profile at this location is clay followed by rock. An earthen Cofferdam of a maximum height of 7 m (23 ft) was made around the periphery of the excavation. Cofferdam embankment and excavation slopes were analysed using the finite element software PLAXIS-2D. Clayey sand was proposed as a filling material during design, but clayey soil was used for the embankment formation during execution. The embankment was built up to 5 m in height in July 2017 and completely submerged due to monsoon. In February 2018, the embankment height was raised to 7 m (23 ft) by filling the soil over the existing submerged embankment. After a few days of completion of the Cofferdam, the inside slope of the embankment had started sliding due to continuous dewatering. The design and execution of deep excavation with earthen Cofferdam for waterfront structures are challenging. The selection of filling material for embankment formation and dewatering techniques plays a vital role in the safety and stability of the embankments. Although all the safety measures were taken during the design, execution challenges are unpredictable.

Keywords Deep excavation · Embankment · Cofferdam · Slope stability · PLAXIS · Dewatering · And sudden drawdown

P. V. Ramana (✉) · M. Balasubramani · K. Bairagi
L & T Construction, Chennai, Tamil Nadu, India
e-mail: ramanapv@lntecc.com

M. Balasubramani
e-mail: mbs@lntecc.com

K. Bairagi
e-mail: kbr@lntecc.com

B. Adimoolam et al. (eds.), *Deep Foundations for Infrastructure Development in India*, Lecture Notes in Civil Engineering 315, https://doi.org/10.1007/978-981-19-8598-0_10

1 Introduction

Cofferdam is a temporary enclosure in or around a body of water that is constructed to allow diversion or damming of an enclosed area. The primary purpose of a Cofferdam is to create a dry working area for a structure to be executed. Cofferdams are typically conventional embankment dams of both earth- and rock-fill, although concrete or some sheet piling also may be used.

1.1 Earthen Cofferdam

This type of Cofferdam is suitable for high heads of water with low velocity. A successful Cofferdam does not need to be completely watertight because some seepage of water into the excavation is usually well tolerated. The water collected is pumped out of the excavation afterwards. The embankment is provided with a minimum freeboard of 1 m (3 ft) to prevent overtopping by waves. This type of cofferdam requires a large base area and is adopted when an area of excavation is very large. Clayey soil is appropriate for construction in the dry season. If constructed in the wet season, the clayey sand or sandy clay fill is the best material.

1.2 Safety

Embankment slope failure may occur in steady-state seepage or sudden drawdown conditions if the soil shear strength parameters are not adequate. Other factors like earthquakes may cause dam failure as referred to in IS: 7894 (1975), but considering construction duration of the structure, this failure is very rare and normally Cofferdams are not designed for earthquakes. Overtopping or overflow of an embankment dam may also cause eventual failure. It is required to provide a minimum freeboard to prevent overtopping failure as per IS: 8826 (1978).

2 Case Study

2.1 Location

The present deep excavation and Cofferdam were located near Mukane Dam on the Aaundha River, Nashik district in the state of Maharashtra in India.

2.2 Physiography

The topography of the area: Uneven, Soil Type: Black cotton soil followed by Basalt rock.

Climate: tropical wet and dry climate.

2.3 Cofferdam and Excavation Details

Natural ground level = varies from average RL + 581 m to RL + 587 m.

High flood level = RL + 594 m.

Average water level = RL + 587 m.

Cofferdam top level = RL + 588 m.

Cofferdam top width = 4 m (13 ft).

Slope of the Cofferdam = 1V:3H (Underwater formation)

The bottom level of the excavation = RL + 575 m.

Excavation slope in clay = 1V:1H, Excavation slope in rock = 3V:1H.

3 Geotechnical Study

3.1 Subsoil Exploration

Four number of boreholes were taken at specified locations to ascertain the information about the sub-soil, its nature and its strength (Table 1). Soil samples were collected at different depths to identify the soil strata and to conduct laboratory tests. The water table was encountered at ground level.

Table 1 Natural ground profile and shear strength parameters

Soil layers	From (m)	To (m)	γ (kN/m^3)	c (kN/m^2)	$\phi°$	UCS (kN/m^2)	Es (kN/m^2)
Clay (CH)	0	−6	17	75	0	–	6600
Clayey sand (SC)	−6	−9	17.5	70	20	–	32,500
Grey basalt	−9	−12	24	–	–	1.385×10^4	20×10^5
Dark basalt	−12	−20	25	–	–	1.533×10^4	30×10^5

3.2 Laboratory Testing

Appropriate laboratory tests were conducted based on IS: 2720 (2006–07). The main tests conducted are listed below.

Tests on Borehole Samples

- Grain size analysis, Atterberg limits (Liquid limit and Plastic limit).
- Natural moisture content, Bulk density.
- Chemical tests on soil and water.
- Un-confined compressive strength on Rock, Triaxial test for determining (c & ϕ).

4 Analysis

Based on the geotechnical investigation report slope stability analyses were done for the embankment and excavation together by PLAXIS-2D.

4.1 Introduction to PLAXIS-2D (2014)

PLAXIS is a finite element package that has been developed, especially for the analysis of deformation and stability in geotechnical engineering projects.

4.2 Analysis with PLAXIS-2D

Cofferdam embankment was analysed considering clayey sand (refer to Table 2) as filling material. Embankment along with excavation was simulated with PLAXIS software for steady-state seepage condition and sudden drawdown condition. Mohr–Coulomb constitutive model and drained shear strength parameters were used in the analysis. The critical slip circle was observed in the downstream side in case of a steady-state condition (Fig. 1), whereas, in sudden drawdown, it was observed in the upstream side (Fig. 2). Minimum factor of safety required for steady-state seepage and sudden drawdown condition was 1.5 and 1.3, respectively, as per IS: 7894. Refer to Table 3 for analysis results (Figs. 3 and 4).

Table 2 Cofferdam filling soil properties

Type of soil	From (m)	To (m)	γ (kN/m^3)	c (kN/m^2)	ϕ°	UCS (kN/m^2)	Es (kN/m^2)
Clayey sand (proposed)	0	7	18	10	30	–	15,000
Clay (actual)	0	7	17	30	10	–	4800

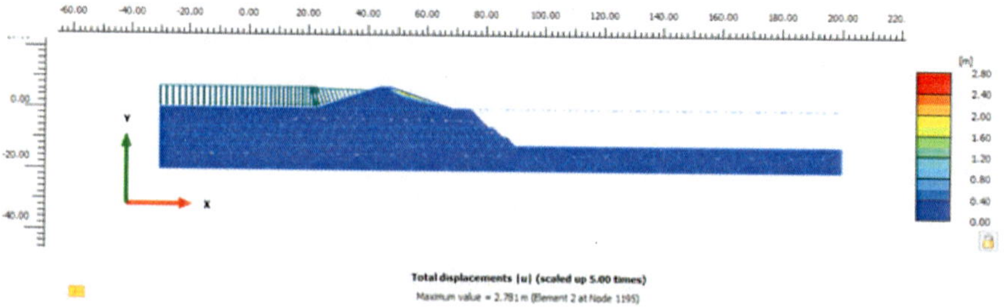

Fig. 1 Critical slip circle in steady-state seepage condition

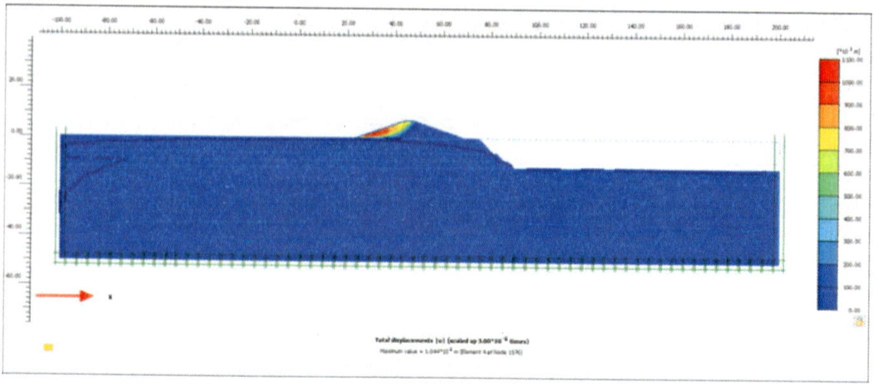

Fig. 2 Critical slip circle in sudden drawdown condition

Table 3 Slope stability analysis results of proposed fill material used for Cofferdam

Type of fill material	The factor of safety in steady-state seepage condition	The factor of safety in sudden drawdown condition
Clayey sand	1.78	3.7

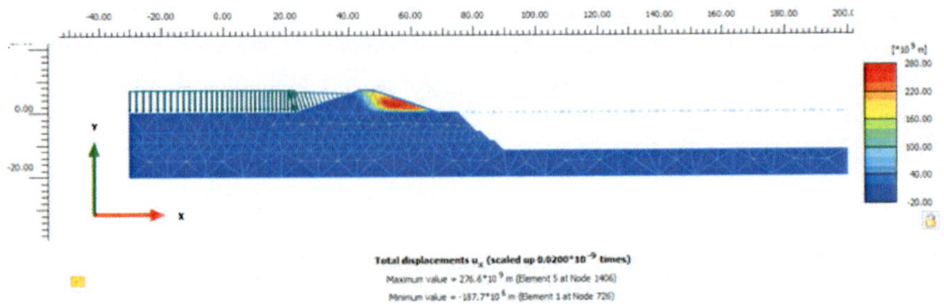

Fig. 3 Critical slip circle in steady-state seepage condition under safety analysis

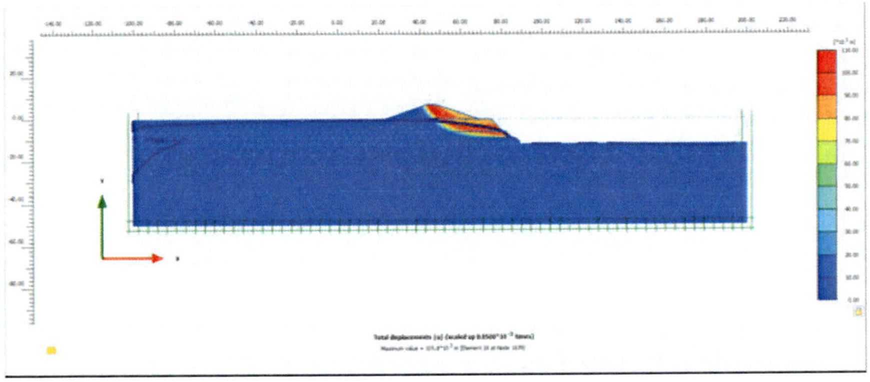

Fig. 4 Critical slip circle in sudden drawdown condition under safety analysis

5 Chronology and Challenges

5.1 Non-Availability of Proposed Filling Material at Nearby Vicinity

Clayey sand proposed for embankment formation and design was also made accordingly. Project location was surrounded by black cotton soil; hence, the Cofferdam was made with available clayey soil. Borrowed soil for embankment formation is indicated in Fig. 5.

5.2 Difficulties in Compaction

The soil was dumped into the water for embankment formation, it was not possible to compact the soil under water. Roller compaction was done once the Cofferdam

Fig. 5 Borrowed soil for embankment formation

top level reached above water level. Therefore, embankment soil under the water was not compacted.

5.3 *Submergence of Cofferdam*

The embankment was built up to 5 m in height in July 2017 and completely submerged due to monsoon. In February 2018, embankment height was raised to 7 m (23 ft) by filling the soil over the existing submerged embankment.

5.4 *Continuous Dewatering*

After completion of the embankment formation, immediately dewatering was started with six numbers of 60 HP pumps and the water level in the inside pool was reduced from RL + 587 m to +581 m within 7 days. On the eighth day inside slope, embankment soil had started sliding slowly downstream (Fig. 6).

5.5 *Checking Stability of the Cofferdam with Actual Soil (Clay) as Fill Material*

Slope stability analysis was performed again with filled soil as clay (Table 2) to understand the probable causes for the failure. Refer to Table 4 for analysis results.

Probable reasons for the failure

(a) The Cofferdam was designed as the proposed fill material such as clayey sand, but the actual Cofferdam was made with clayey soil.
(b) The embankment was constructed in two seasons, in the first season, it was submerged and in the next season, embankment was made over the existing

Fig. 6 Soil is prone to slide at downstream

Table 4 Slope stability analysis results of proposed fill material for Cofferdam

Type of fill material	The factor of safety in steady-state seepage condition	The factor of safety in sudden drawdown condition
Clay	3.36	4.8

submerged embankment. The soil became very soft due to submergence and a weak interface had developed between the submerged embankment and dry embankment.

(c) Continuous dewatering resulted in sudden drawdown conditions and excessive pore water pressure developed in the embankment, finally led to downstream side soil sliding.

Remedy

(a) Dewatering was stopped immediately after the sliding was observed.
(b) The embankment was strengthened by increasing the top width from 4 m (13 ft) to 7 m (23 ft).
(c) Proper Compaction was done by Roller (Fig. 7).
(d) Downstream side soil was prone to slide was removed completely (Fig. 8).
(e) The excavation was started after 15 days, as the water level reduced on the upstream side due to the summer season has begun.
(f) The project was completed in the same season and Cofferdam was removed.

Fig. 7 Roller compaction of soil filled on upstream side

Fig. 8 Removing the soil, prone to slide

6 Summary and Conclusions

This paper elaborated on the design, analysis method, and execution challenges of deep excavations for waterfront structures. Although proper care was taken during design, execution challenges are unpredictable. Following are the learning experiences:

i. Cofferdam shall be constructed up to the desired level in one season to avoid the maximum possible submergence.
ii. While designing the earthen Cofferdam, designer should check the availability of filling material nearby project sites to avoid the changes from design to execution.
iii. Continuous dewatering causes sudden drawdown conditions in embankments made with clayey soils which may lead to slip circle failure.
iv. The selection of filling material for Cofferdam, compaction procedure and dewatering techniques play a vital role in the completion of the project on time.
v. Also, it is advisable to provide HDPE sheet on upstream side of the embankment to control seepage during excavation.

References

IS: 7894 (1975) Code of practice for stability analysis of earth dams. Bureau of Indian Standards, New Delhi, India

IS: 8826 (1978) Guidelines for the design of large earth and rockfill dams. Bureau of Indian Standards, New Delhi, India

IS: 2720 (2006–07) Methods of tests for soils. Bureau of Indian Standards, New Delhi, India

PLAXIS (2014) PLAXIS-2D: reference manual, Annual. Plaxis BV, Delft, The Netherlands

Confined Space and Inner-City Projects—Future Challenge and Opportunity for Diaphragm Walling

Franz-Werner Gerressen and Alexander Blatt

Abstract Diaphragm walls are known as underground structural elements commonly used as retention systems for excavation pits and shafts and permanent foundation walls or elements. One can anticipate that global urbanization and increasing demands on environmental considerations will need to be accommodated in underground space in the future. These trends show an increasing requirement for diaphragm walling in even more complex conditions, especially in inner-city applications with limited space. Complex conditions in terms of space limitations, especially for inner-city jobsites, require specifically adapted solutions for slurry, spoil, reinforcement and concrete handling and the related logistics to ensure smooth production. Furthermore, one focus will be given to the QA/QC topics of the production process. Real-time installation control, data transfer and reporting systems become more and more important. Therefore, the paper will describe the construction method and the sequence of activities required for the construction of diaphragm wall systems. It will also describe the main equipment that will be needed to execute these works under various conditions. In addition to the general description of the system and the required equipment, the paper will show some site examples for infrastructure projects with their specific solutions when working under space restrictions.

Keywords Diaphragm wall · Trench cutter

F.-W. Gerressen (✉) · A. Blatt
BAUER Maschinen GmbH, Schrobenhausen, Germany
e-mail: franz-werner.gerressen@bauer.de

A. Blatt
e-mail: Alexander.Blatt@bauer.de

B. Adimoolam et al. (eds.), *Deep Foundations for Infrastructure Development in India*, Lecture Notes in Civil Engineering 315, https://doi.org/10.1007/978-981-19-8598-0_11

1 Increasing Demands for Complexity

Within the last decades, the development of diaphragm walling led to a wide range of applications for that technique. To date, over 50% (around 4 billion people) of the world's population already lives in cities and urban areas, with an increasing trend (Ritchie 2018).

The globalized trend of growing cities and expansion of megacities will call for new infrastructure in the future. The limits for new underground metro lines, water waste systems, excavation pits, etc. will be further pushed, as existing structures hinder a new design of areal expansion. Deep excavation under such scenarios requires technology, which is able to reach great depth and penetrate difficult soil and hard rock conditions. The development of the diaphragm wall technology already combines projections to solve these challenges and will underlie a constant advancement in the future.

The trends lead to new challenges for equipment and site logistics, as the space within existing megacities is very expensive and highly limited. Nevertheless, experiences from past jobsites and developments of equipment suppliers show possibilities to overcome the restrictions (Figs. 1 and 2).

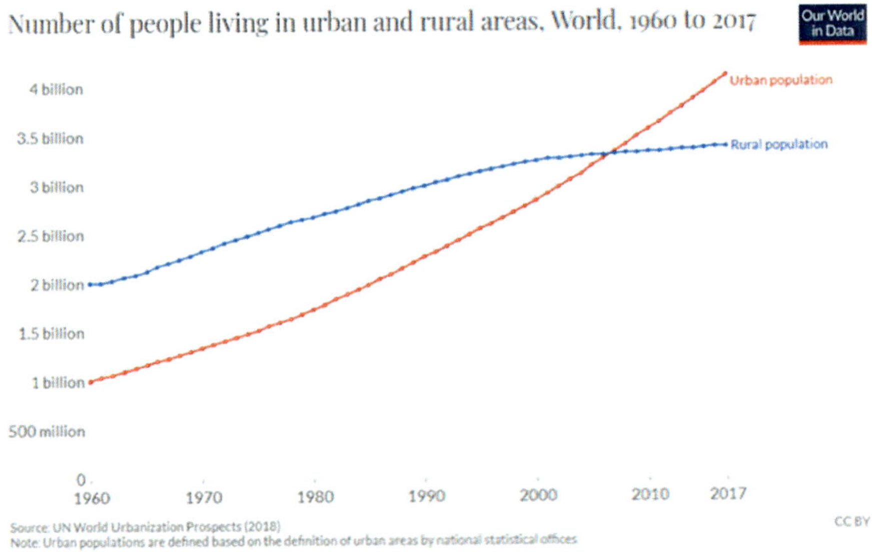

Fig. 1 Development of population in cities (Ritchie 2018)

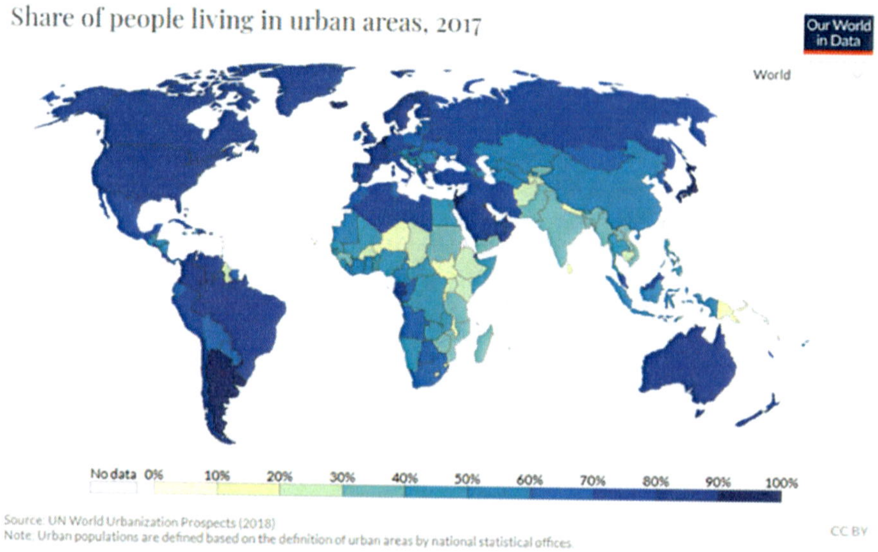

Fig. 2 Map of urban areas (Ritchie 2018)

2 The Working Principle; Focusing on Confined Situations

2.1 *Planning and Adjustment*

Confined space sites especially need good planning, prior to the start of mobilization, and logistics to minimize downtimes during the construction process.

- Develop work plan
- Develop schedule
- Calculate material requisition
- Develop drawings and layouts
- Plan and perform site mobilization.

During the development of the work plan, it needs to be ensured that the iterating steps of the construction process do not interfere with each other. Therefore, changes should be anticipated prior to the start of the site. This minimizes or eliminates delays and cost escalation.

This plan has to be made for each site individually and cannot be generally created. An interaction chart, which shows the relations of important points, is shown in Fig. 3.

The dynamic change on a jobsite within confined space conditions leads to influences of different processes and needs adjustment according to the development of the site. The appearance of a site can change during its advance, but it should be preconceived that major relocations of material should be limited to a minimum and has to be planned accordingly. As examples, the following dynamic guidelines should be considered during the planning and construction stage:

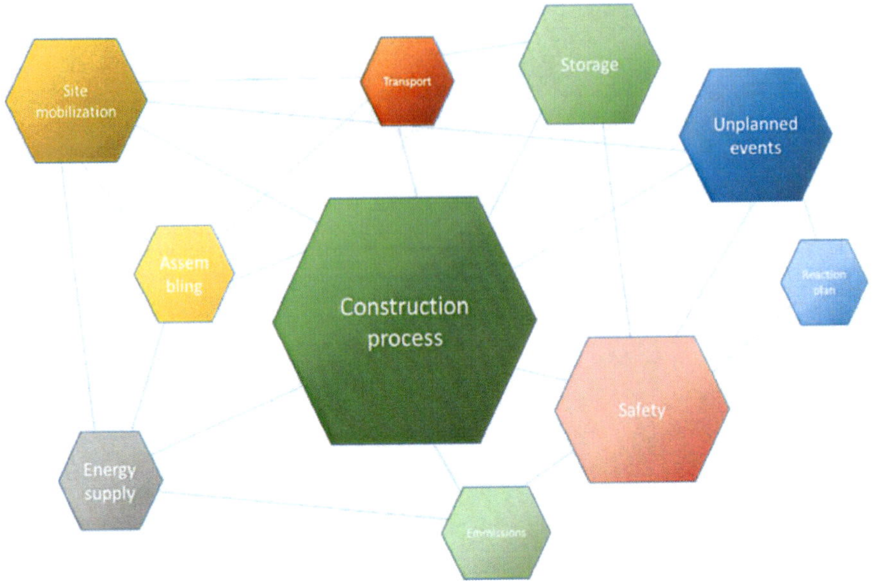

Fig. 3 Interactions of construction process

Plan…

- Choice of suitable method and equipment
- Available space and define zones for storage, equipment, etc.
- Movement of storage areas and equipment
- Alignment of supply and utility lines
- Alignment of roads and walkways
- Handling of material.

Perform…

- Changes according to the work plan
- Processes by following restrictions and safety procedures.

Record…

- All changes and procedures
- Delays
- Constant comparison of times with the planned schedule.

Adjust…

- Work plan, if necessary
- Layouts, procedures according to unplanned events or changed schedule.

2.2 Construction Sequence

The working steps follow a pre-defined sequence. The main steps are shown in Fig. 4.

Working steps:

- Site preparation
- Guide wall construction
- Trench pre-excavation
- Panel excavation
- Panel cleaning (de-sanding)
- Reinforcement installation (for retaining walls)
- Concreting.

In confined conditions, the use of special auxiliary equipment (e.g. for the installation of reinforcement cages) might become necessary.

To meet tight schedules and avoid the cost explosion of a project, the quality of the product is of utmost importance for success. Real-time operation and quality control of state of the art and fast and effective data transfer and analysis will become more important in the future.

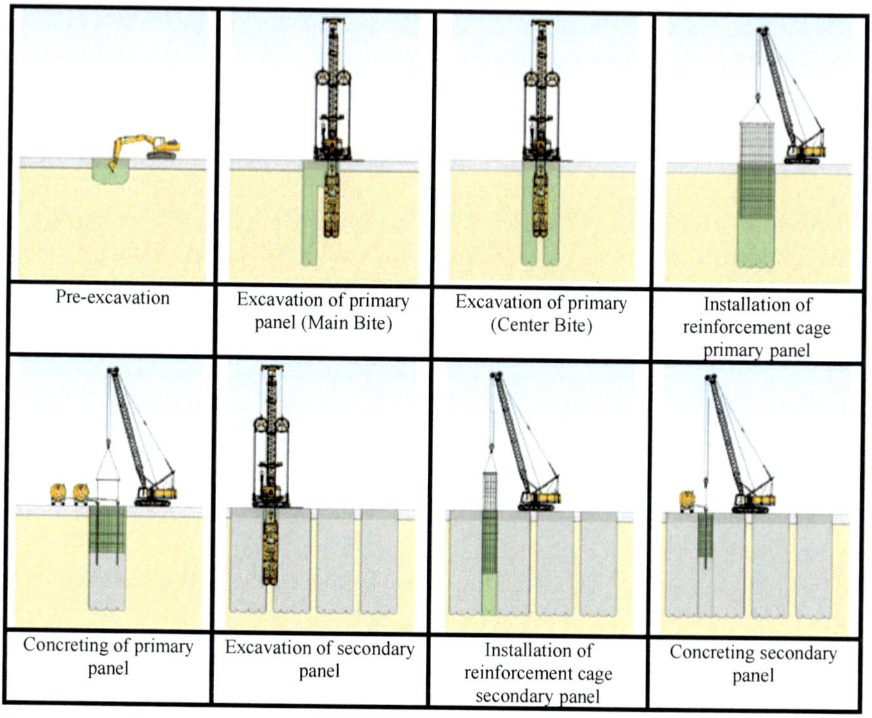

Fig. 4 Construction sequence

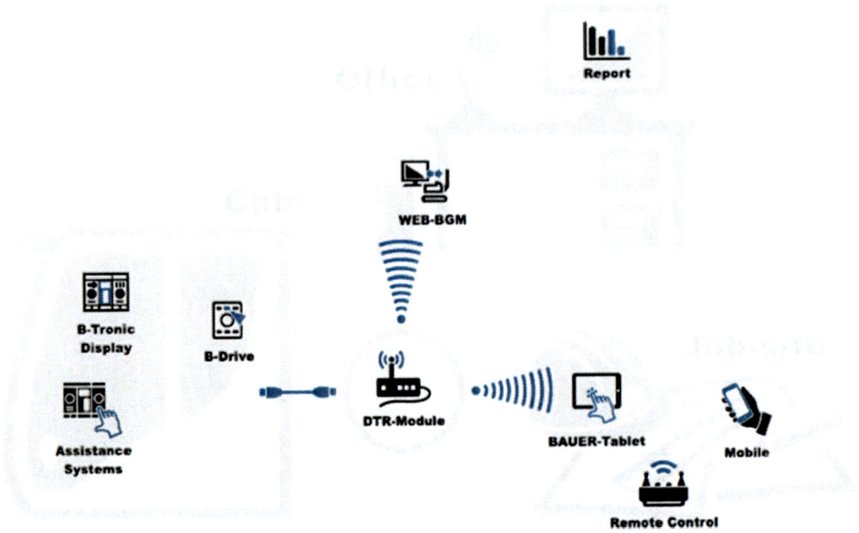

Fig. 5 Data transfer solution

Today, operators are able to receive a visualization of the tool's movement on the ground and control its parameters on a single screen. That makes controls more and more intuitive and ensures high quality.

Online data transfer equipment, nowadays have the possibility to communicate with each other, e.g. to regulate the concrete flow. Furthermore, with this technology, the transmission of machine data, like fuel consumption, engine hours and the transfer of production data or a live view of the operator's monitor on any computer is possible. Afterwards, an effective evaluation and assessment are possible. By visualizing, the data can be used as a basis for planning of future sites. Here, also a progression for a better calculation over time, to predict site developments is imaginable.

An overview of an exemplary communication set up is shown in Fig. 5.

3 Equipment and Tools

For inner-city and projects of confined space, equipment suppliers try to find innovative solutions for different restrictions. Special base carriers as shown in Fig. 6 allow applications, e.g. with height restrictions under existing structures or within tunnels. Special tools combine proven technique with compact modules for restricted conditions.

The height of equipment, for example, can be reduced to around 5–6 m from around 20–25 m, compared to a standard equipment that is capable of reaching the same depth.

Fig. 6 Low headroom cutter

Confined space situations (except height restrictions) do not mandatorily need specialized equipment and specially designed tools. With sufficient planning, for example, turning devices, standard equipment can be used. Jobsites examples can be found all around the world, e.g. installation of cut-off walls (COW) from dam crests of narrow width.

In the future, development in terms of electrification, reduction in emissions and even more powerful machines can be expected. This will further adapt the technology to inner-city applications as environmental suitability becomes more important.

4 Confined Jobsites

4.1 Hung Hom Station, Hong Kong, 2013

In 2013, the Hung Hom Station in Hong Kong was constructed. Figure 7 shows the jobsite layout and the project specifications showed height restrictions of 5.8–6.5 m (see Fig. 8), because of existing structures. In addition, many bearing columns hindered the mobility within the area. For that reason, two special carriers (low headroom models) were in use to enter the site and install a total of 261 panels with a depth of up to 55 m. As the wall, 1200 mm in width, had to key into Granite of Grade III, Round Shank Chisel wheels as shown in Fig. 9 were used. This job is an example of difficult site logistics with the combination of challenging hard soil conditions.

4.2 Grand Metro Paris, France, 2018

The Grand Metro Paris project started in the year 2016 and is known as one of the biggest urban projects in the world (2020). In fact, it currently is the largest

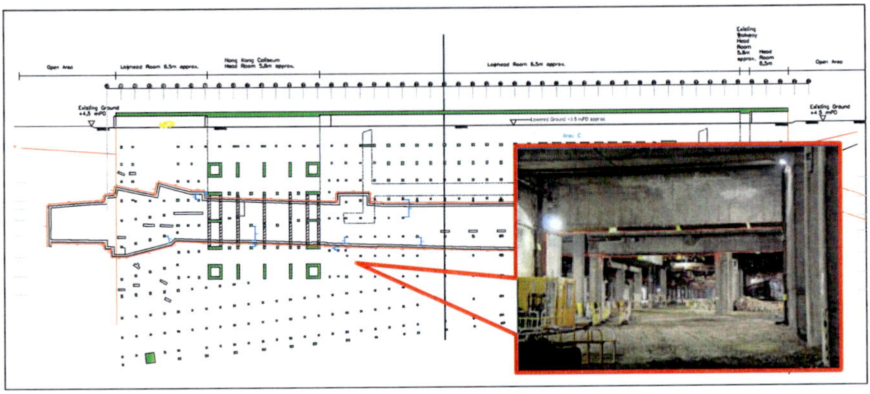

Fig. 7 Jobsite layout

Fig. 8 Low headroom situation

infrastructure project in Europe, with its expanse all over the Parisian area. The large scope of work comprises 4 additional metro lines, 200 km of new railway lines, of which 90% is underground, and 68 newly designed metro stations, which connect the city centre of Paris with outer urban areas (2020). At numerous sites, working in restricted conditions is inevitable. Two site examples from line 15 (Ligne 15) are shown here.

Under an existing bridge of 6.5 m in height a special carrier, as shown in Fig. 10, had to be used to penetrate into hard soil conditions.

Fig. 9 RSC cutting wheels

Fig. 10 Low headroom cutter

In the second site along line 15, the width of the working area for logistics, construction and storage was limited to a width of 20 m. Therefore, a compact mini cutter (Fig. 11) penetrated around 20 m of limestone to reach final depths greater than 60 m.

4.3 Dam Recovery Rosshaupten, Germany 2018

A dam recovery project in Rosshaupten, Germany, was completed in 2019. Overall, around 227 m, 14,000 m^2 of cut-off wall (COW) had to be constructed, to stop water seepage in the karst formation below the dam and avoid erosion effects in the dam body as shown in Figs. 12 and 13.

The COW was installed from the dam crest, which had a width of 11 m only. In addition, the axis of the sealing wall was not aligned in the centre of the crest, which required a relative turning of the excavation tool to the base machine (see Fig. 14).

Fig. 11 Mini cutter

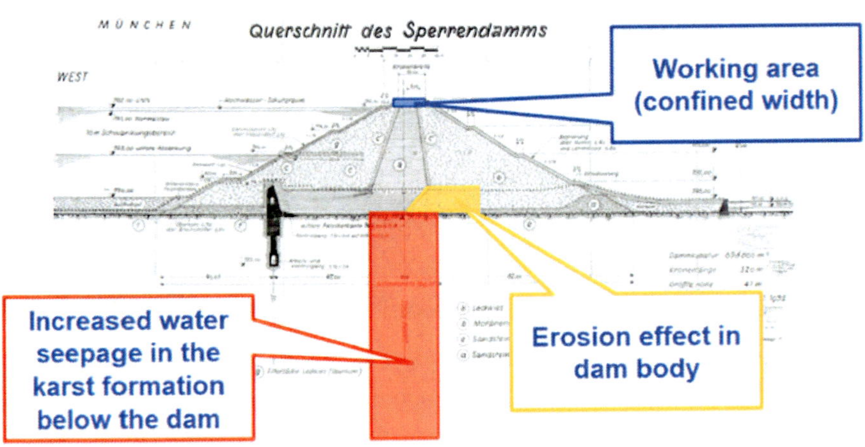

Fig. 12 Cross section of dam

Working in a rectangular position of tool and base carrier was impossible due to the small area. This required a good work plan and pre-defined cycles for the movement of the equipment, which had standard dimensions. A grab unit and a cutter unit worked in combination to excavate the panels along the axis of the wall to a depth of up to 68 m (see Fig. 15). Around 41 m of the dam material was excavated by grab before the cutter embedded and sealed the permeable karst formation.

As part of a recreation area, the execution of the work took place during the winter months (see Fig. 16). This led to further challenges and procedures to succeed in the site, e.g. heating the slurry storage area within a constructed tent.

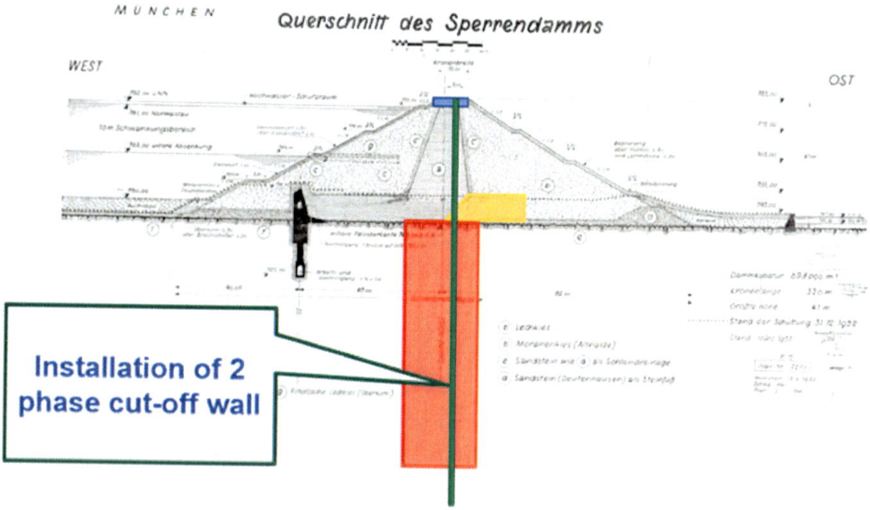

Fig. 13 Alignment of COW within dam cross section

Fig. 14 Standard cutting unit with turning device

Fig. 15 Combination of grab and cutter

Fig. 16 Winter conditions

5 Conclusion

With an increasing number of people moving into urban and inner-city areas, the demand for new infrastructure will also soar in the future. Existing buildings and structures, as well as high prices for areal spread, set restrictions and demand for the construction of these new infrastructure facilities. In addition, environmental aspects will become more important and will raise the question of new strategies. The impact of these demands will further influence the construction process and the design of equipment.

The importance to provide solutions for confined space jobsites with challenging soil conditions already raises difficulties for logistics, storage and methods as well as equipment to solve the restrictions. Today, suppliers already provide compact machines of high performance. Especially, diaphragm wall method already shows the ability to work in nearly all soil and rock conditions and find solutions for confined space situations.

In the future, further developments of powerful equipment and methods to work within confined spaces are expectable. Furthermore, efforts to reduce noise and fuel emissions will take place to prepare the equipment for a wider spread of inner-city applications.

References

https://www.societedugrandparis.fr/ (2020)
Ritchie H (2018) Urbanization. On https://ourworldindata.org/urbanization

Effects of Change in the Support System on Temporary Secant Pile Wall

C. Anburaj and A. Srinivas

Abstract Deep excavation with a support system is required to construct various parts of underground structures like shafts, stations, entry structures, etc. These structures have to be constructed using permanent or temporary embedded retaining walls with a support system. The selection of a suitable type of retaining wall will depend upon the geological condition present in the particular location, time, cost, available equipment, etc. In Bangalore Metro Rail Project, a secant pile wall was adopted as the temporary earth retaining system with the depth of excavation of about 20 m. It was designed initially based on bottom-up construction methodology with 3 levels of struts and 3 levels of anchors, but at a later stage, due to time and other construction-related issues, it was decided to change the configuration to 6 levels of struts. Generally, in deep excavations, all the underground structures should be designed and checked for the critical forces from both permanent stage and construction stages. Since secant piles are used as a temporary retaining wall, only construction stage analysis is carried out to get the governing forces and deformation. In construction stage analysis, soil layers are defined with boundary conditions, and the surcharge during construction and surcharge from the actual building near the secant pile are considered. This paper discusses the effect of change in the support system from anchors to struts and how these changes in the support system affect the behaviour of the secant pile and subsequently adjacent buildings present in the influence zone of the excavation. As a result, in changing from anchors to struts, the wall displacement, strut forces and ground movement on the adjacent ground and buildings increase. The increase in deflection, ground movements and strut forces occurs during backfilling sequence of the underground station excavation. During backfilling, struts have to be removed and this imparts a higher magnitude in deflection, strut forces and ground movements.

Keywords Support system · Anchors · Struts · Ground movement · Settlement

C. Anburaj (✉) · A. Srinivas
L&T Construction, Chennai, Tamil Nadu, India
e-mail: anburaj-c@lntecc.com

A. Srinivas
e-mail: a-srinivas@lntecc.com

B. Adimoolam et al. (eds.), *Deep Foundations for Infrastructure Development in India*, Lecture Notes in Civil Engineering 315, https://doi.org/10.1007/978-981-19-8598-0_12

1 Introduction

Construction of underground metro stations in most cities is growing at a faster rate in our country. The general methodology adopted in metro stations seems to be familiar, but a lot of risks are involved due to variable geological conditions. Furthermore, in the case of structures below ground, the close contact of the surrounding ground with the structure, soil–structure interaction plays a prominent role in design. As a result, there is a close interrelationship between the method and sequence of construction of the structure and its design. In other words, the design of underground metro station (cut-and-cover structures) cannot be treated separately from their construction. Therefore, the design and construction of such structures, particularly in a highly constrained urban environment and difficult ground and groundwater conditions, will present some of the greatest challenges to both the designer as well as the constructor. Therefore, for the construction of the underground metro station, a temporary retaining structure is adopted. Considering ground conditions, groundwater conditions, excavation depth, construction equipment, time and cost of construction, secant piles are proposed as temporary earth retaining structures. On observing the site conditions, the Bottom-up method of construction is followed in one of the underground metro stations in the Bangalore Metro Rail project. To retain the soil, struts and anchors are used to provide lateral support to the secant pile walls initially. Over time, due to construction issues, anchors are replaced with struts. Due to the change in the support system, the behaviour of the secant piles and the ground response will change because of the fact that the stiffness of the whole system wall will change while replacing it. Therefore, to analyze the behaviour of the secant pile wall, two support systems are considered. They are (i) 3 strut 3 anchor configuration and (ii) 6 strut configuration. This paper presents the effects on the behaviour of the secant pile wall and also the ground response when anchors are replaced by struts. The response of the adjacent buildings when anchors are changed to struts is also discussed. The changes in strut forces after changing the support system are examined and the results are presented as a part of this study.

2 Site Geology

The reduced level (RL) of the ground varies between +904.500 m and +907.838 m. The main rock formation in the Bangalore area is granitic gneiss. The granitic gneisses are mainly of migmatitic type, highly banded varying in composition from granite to diorite. Grade IV rock is encountered at about 26 m below the ground level. The type of rock generally observed is Granitic Gneiss. Rock is encountered at some locations and is not uniform throughout the station. The water table varies from 1.2 to 18 m below the ground surface based on the investigation. As per the CGWB survey, most of the locations in Bangalore have deeper water level ranging from 10 to 20 m below ground level. Hence, the fluctuation in the water table is due to the effect of the

Table 1 Geotechnical properties of the sub-soil profile

Depth (m)	Strata type	γ (kN/m^3)	E' (MPa)	c' (kPa)	ϕ'
0–2.5	Filled up	18	10	–	28°
2.5–7.5	Clay & silt of low plasticity	18.5	11 + 3.8z	26	24°
7.5–16.5		18.5	30 + 3.8z	46	25°
16.5–21	SM	18.5	64.2 + 3.8z	–	38°
21–26	SC	18.5	81.3 + 3.8z	15	23°
>26	Soft rock	18.5	136	–	40°

perched ground water table and frequent rainfall noticed during the investigation. From the ground profile, a 2 m thick layer of aquiclude retains the water and behaves as an impermeable layer. The generalized profile of the station considered is shown in Table 1.

3 Construction Methodology

It is planned to construct the underground metro station by the Bottom-up method of construction. The bottom-up method is ideally suited for the particular location because of the following reasons:

- There are no major services or utilities in the vicinity of the construction site
- Very less disturbance to the on-road traffic
- Ground movements anticipated are low and groundwater pressures are not high
- Adjacent buildings and structures close to excavation are very less in number.

In the Bottom-up method, as a first step, a capping beam is constructed which helps in connecting the secant pile wall intact from the top to the bottom of the excavation and also prevents the out of plane and lateral movement of the wall when exposed during excavation. After the capping beam acquires the required strength, a proper dewatering system is adopted to lower the water table within the station so that excavation can be carried out in a dry state. This is usually done for any excavation activity because the presence of groundwater will make excavation tedious. Hence, before progressing to every level or stage of excavation, the groundwater table has to be lowered within the perimeter of the excavation. As the excavation progresses downwards, struts or anchors are used to provide lateral support to the walls. After reaching the final or desired excavation level, construction of the structure is then carried out. The bottom slab is constructed first, which progresses upwards in a conventional manner. As the construction of the structure advances upwards and when the elements of the structure which form the permanent lateral supports are in place and operational, each corresponding stage of temporary horizontal bracing is sequentially released and removed ensuring adequate safety measures. The back-filling and surface reinstatement are finally carried out only after the roof slab and the associated support structure are in place and have attained the requisite strength to sustain the imposed loads.

3.1 Supports System

For the construction of the metro station, the depth of excavation is about 20 m below the ground level. Hence, to provide lateral support to the secant pile wall, three levels of struts and three levels of anchors are used during design. The details of the strut and anchor used during the excavation are mentioned in Table 2.

The strut and anchor levels used to retain the soil for 20 m depth are presented in Table 3. The combination of struts and anchors is chosen in such a way that it causes less hindrance to construction activity and requires optimum reinforcement for the secant piles.

However, in a later stage, due to construction issues, the last three levels of the anchor are replaced by struts and the levels are slightly altered. The fourth, fifth and sixth level of struts is raised by 1.58 m, 2 m and 1 m, respectively, from their initial anchor position.

3.2 Construction Sequence

As mentioned earlier, the bottom-up method is adopted for construction. The typical stages of the construction sequence are as follows:

Table 2 Details of strut and anchors

Type of support	Sectional details	Young's modulus in kPa
Struts	2UB 610 × 229 × 125.1	2×10^8
Anchors	7-strand of 15.7 mm diameter	2×10^8

Table 3 Levels of struts and anchors

Levels of support	Depth from existing ground level (m)
First level strut	2.5
Second level strut	6
Third level strut	9
Fourth level anchor	12.5
Fifth level anchor	15.5
Sixth level anchor	17.5

- Installation of secant piles after capping beam gains enough strength.
- Dewatering system is installed prior to excavation. Water table is maintained 1 m below excavation.
- Excavate 0.75 m below the first level strut and install the first level of strut.
- Excavation is carried out in a similar manner up to the installation of sixth level strut.
- After reaching the final excavation level, the base slab is constructed against secant pile wall with proper waterproofing layers.
- After casting the base slab, the station wall is constructed with internal elements after the removal of corresponding struts.
- Construct the concourse and roof slab in a similar way after removing the respective struts and anchors. Backfilling is done between the sides of the secant pile and permanent walls.
- Reinstating the ground to the original level by backfilling.

4 Finite Element Analysis

In general, finite element analysis is used to predict the behaviour and response of the temporary or permanent retaining structures in a robust manner. Even though the secant piles are temporary structures, the analysis is carried out for the construction stage and permanent stage, i.e., undrained parameters of soil for the temporary stage and drained parameters for the permanent stage. Therefore, Secant piles, struts and anchors are designed for the critical case scenario. The finite element analyses were carried out using WALLAP software to predict the deflection of secant pile wall, strut and anchor forces. The properties of secant piles, struts and anchors used in the analysis are given in Table 4.

The construction sequence as said above is exactly modelled in WALLAP and the distribution of bending moments, shear forces and deflection of the secant pile is obtained for both the cases, i.e., with anchors and anchors replaced by struts. The change in the strut forces in the first three levels of struts, due to the replacement of anchors, is also observed. Similarly, the ground movements due to changes in the support systems are also predicted using Clough and O'Rourke (1990) method. The forces on struts, ground movements and tensile strain induced in the adjacent buildings due to changes in the support system are presented in the subsequent sections.

Table 4 Material properties used in WALLAP

Element type	I (m^4/m)	Cross-sectional area (m^2)	E (kPa)	Spacing (m)
Secant pile	0.014893	0.5026	2.74×10^7	1.35
Struts	–	0.03186	2×10^8	9.0
Anchors	–	0.000973	2×10^8	2.7

5 Estimation of Ground Movements and Tensile Strain of Adjacent Buildings

Ground movements can be computed by the principles given by Bowles (1990) and Clough and O'Rourke (1990) depending on the type of soil. Based on several case histories, Clough and O'Rourke (1990) suggested that the settlement profile is triangular for excavation in sandy soil or stiff clay. The maximum ground surface settlement will occur just behind the wall. The non-dimensional profiles are given in Fig. 1. It shows that the corresponding settlement extends to about $2H_e$ and $3H_e$ for sandy soil and stiff to very hard clays, respectively, where H_e is the influence depth.

As the excavation progresses, the lateral pressures imposed by the ground behind the wall would induce wall deflections into the excavation. This would result in vertical and lateral displacements of the ground adjacent to the retaining wall. In principle, the magnitude and extent of this ground movement are a function of the retention system type, the adopted construction methodology and the properties of the soil and/or rock materials. The depth of influence (H_e) is considered depending on the depth of the secant pile wall and the depth of the excavation. The geology of Bangalore is predominantly mixed soil condition comprises mainly of mixtures of silty sand and clays with low to high plasticity and compressibility. The maximum deflection on the ground adjacent to the secant pile due to excavation at the launching shaft can be estimated with the deflection profile of the secant pile. As the strata are generally mixed, the typical settlement profile just behind the secant pile wall as shown in Fig. 2.

The ground settlement curve is taken as "second degree exponential curve" as suggested by Bowles (1990) where the maximum ground settlement occurs just behind the wall. Bowles (1990) suggested a procedure to estimate the excavation-induced ground surface settlements using the following relations:

$$\delta_v = \delta_{vm}\left(\frac{l_x}{D}\right)^2 \tag{1}$$

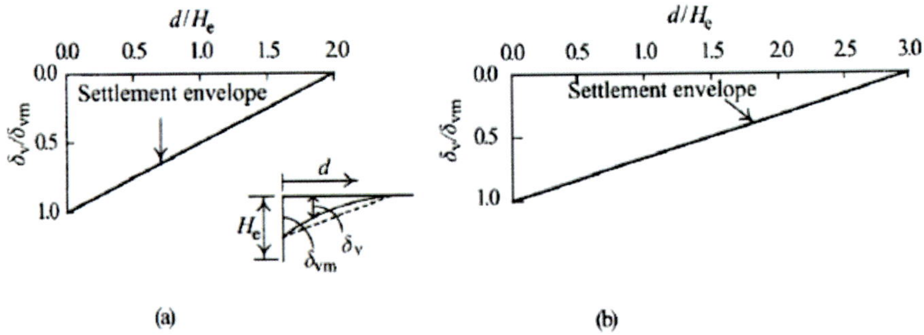

Fig. 1 Dimensionless Settlement profiles adjacent to Excavation **a** sandy soil **b** stiff clays (Clough and O'Rourke 1990)

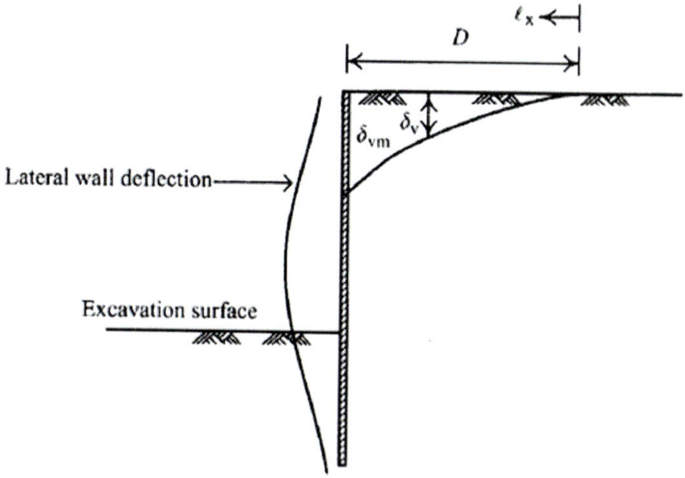

Fig. 2 Estimation of ground settlement (Bowles 1990)

where δ_v = settlement at a distance of $D - l_x$, δ_{vm} = maximum ground surface settlement, l_x = Distance from a point at a distance D from the wall and D is the influence range of ground surface settlement.

The maximum ground surface settlement δ_{vm}, is estimated from the following equation:

$$\delta_{vm} = \frac{2A}{D} \tag{2}$$

where A is the area of the lateral wall deflection. The lateral deflection of the secant pile wall is obtained using finite element software, WALLAP.

5.1 *Limiting Tensile Strain Method*

Burland and Wroth (1974) and Burland et al. (1977) applied the concept of limiting tensile strain to elastic beam theory to study the relation between building deformation and the onset of cracking. Although modelling a building as an elastic beam is a simplification, it was found that predictions from this model were in good agreement with case records of damaged and undamaged buildings. Furthermore, this simple approach demonstrates the mechanisms which control the onset of cracking within a structure.

The elastic beam in their model is described by a width, B and a height, H (see Fig. 3). The figure shows two extreme modes of deformation: In bending, cracking is caused by direct tensile strain, while in shear diagonal, cracks appear, caused by diagonal tensile strains. For a centrally loaded beam subjected to both shear and bending deformation, the total central deflection is given by Timoshenko (1955)

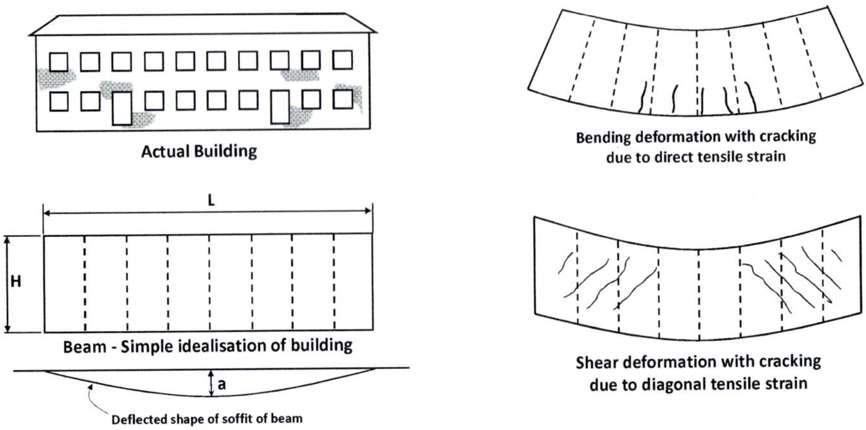

Fig. 3 Cracking of a simple beam in different modes of deformation (Burland and Wroth 1974)

$$\Delta = \frac{PB^3}{48EI}\left(1 + \frac{18EI}{B^2HG}\right) \tag{3}$$

where E—Young's modulus

G—Shear modulus

P—Point load, which is applied at the centre of the beam.

For an isotropic elastic material, E/G = 2(1 + ν). Assuming a Poisson's ratio of ν = 0.3, one obtains E/G = 2.6. In the case where the neutral axis is in the middle of the beam, Burland and Wroth (1974) expressed the above equation in terms of deflection ratio and the maximum extreme fibre strain ε_b

$$\frac{\Delta}{L} = \frac{\varepsilon_v}{3}\frac{\left(1 + 6\frac{E}{G}\frac{H^2}{L^2}\right)}{\left(1 + 4\frac{E}{G}\frac{H^2}{L^2}\right)} \tag{4}$$

I = $H^3/12$ for sagging zone (t = H/2 in sagging zone).

I = $H^3/3$ for hogging zone (t = H in hogging zone).

$$\varepsilon_b = \frac{(\Delta/L)}{\left(\frac{L}{12t} + \frac{3EI}{2tLGH}\right)} \tag{5}$$

$$\text{Diagonal strain } \varepsilon_d = \frac{(\Delta/L)}{\left(1 + \frac{GHL^2}{18EI}\right)} \tag{6}$$

The horizontal strain ε_h is calculated as mentioned in Fig. 4 and is given by the following expression:

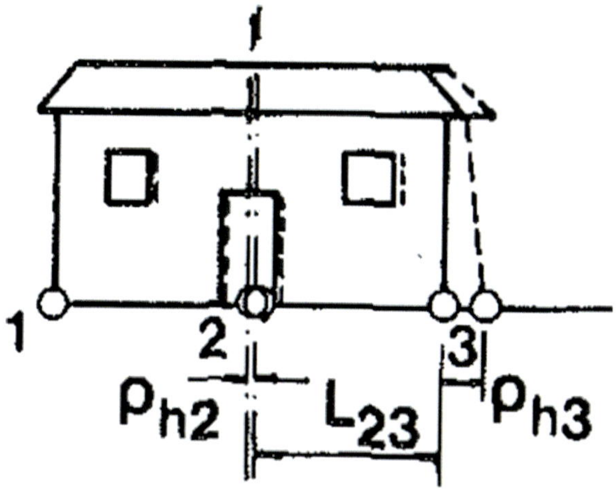

Fig. 4 Typical lateral movement of the buildings near excavation (Boscardin and Cording 1989)

$$\varepsilon_h = \frac{\Delta_{h1} - \Delta_{h2}}{L} \tag{7}$$

where Δ_{h1} and Δ_{h2} are the lateral movement of the building at the ground surface and L is the length of the building in the influence zone of excavation. The limiting tensile strain in the building (ε_t) is the maximum combined horizontal bending strain or combined horizontal and diagonal shear strain.

$$\varepsilon_t = Maximum\ of\ \left\{ (\varepsilon_h + \varepsilon_b), \frac{1-\nu}{2}\varepsilon_h + \sqrt{\left(\frac{\varepsilon_h(1+\nu)}{2}\right)^2 + \varepsilon_d^2} \right\} \tag{8}$$

In the above equation, when Poisson's ratio (ν) is taken as 0.3, then the resulting equation is given below

$$\varepsilon_t = Maximum\ of\ \left\{ (\varepsilon_h + \varepsilon_b), 0.35\varepsilon_h + \sqrt{(0.65\varepsilon_h)^2 + \varepsilon_d^2} \right\} \tag{9}$$

6 Results and Discussion

Results of numerical analysis from WALLAP considering the three levels of struts and three levels of anchors with six levels of struts show that the deflection of the secant pile is high when six levels of struts are used even though the stiffness of strut is higher than that of anchors. This is due to the fact that, while removing or backfilling sequence, i.e., after construction of the base slab, the internal horizontal struts have to be removed to proceed with the construction further. In this scenario, the secant pile wall is subjected to a maximum cantilever span. However, when anchors are there, these anchors will not hinder the construction of internal elements. Hence, these anchors need not be removed during backfilling sequence, but de-stressing is required, i.e., load at the anchor head will be removed during backfilling and not the entire anchor. The deflection profile of the secant pile for both the cases is presented in Fig. 5.

The maximum deflection of the secant pile is only in six strut configuration which yields 7 mm more than the deflection in strut and anchor configuration. However, the deflection pattern of the secant pile wall seems to be identical. The increase in deflection due to the change in the support system is about 13% from the original configuration.

Due to the replacement of anchors by struts, the forces estimated on the first three levels of struts increase. This is because during backfilling, the unsupported length of secant piles would have caused the struts at the top to consume more force. In addition, the struts are internal compressive members, and when secant piles move inwards, the forces on the struts will increase. The variation in strut forces due to change in the support system is mentioned in Table 5.

From Table 5, it is observed that the first level strut is subject to very less increment when compared to the remaining two levels. From this, we can conclude that due to the change in supports at lower levels, there will not be much variation for the supports at the top level. Nevertheless, the struts, which are above the anchors, are experiencing a significant increase in the strut forces. As suggested by Bowles (1990),

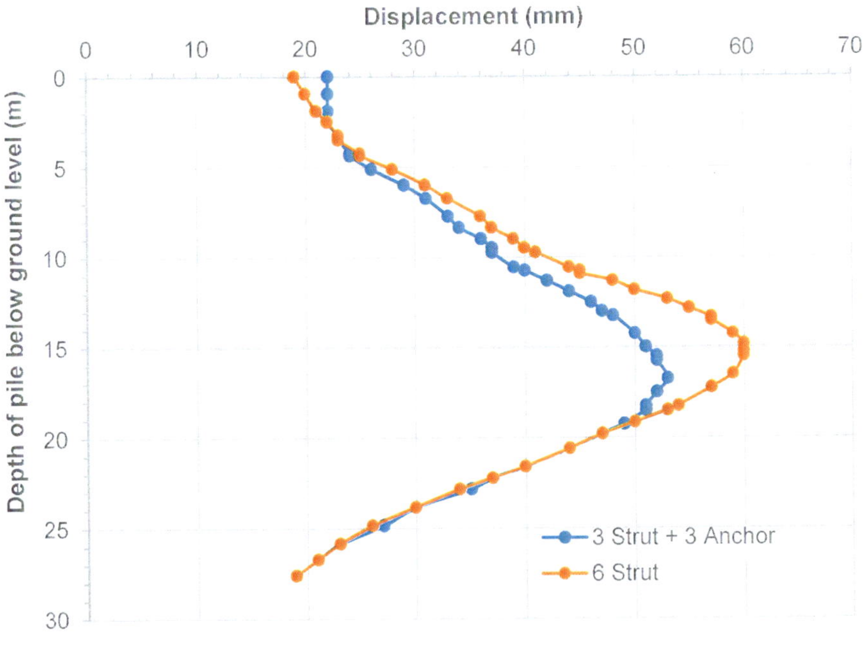

Fig. 5 Deflection profile of secant pile wall

Table 5 Increase in strut forces due to change in support system

Strut levels	Increase in strut forces (%)
1st level strut	5
2nd level strut	10
3rd level strut	17

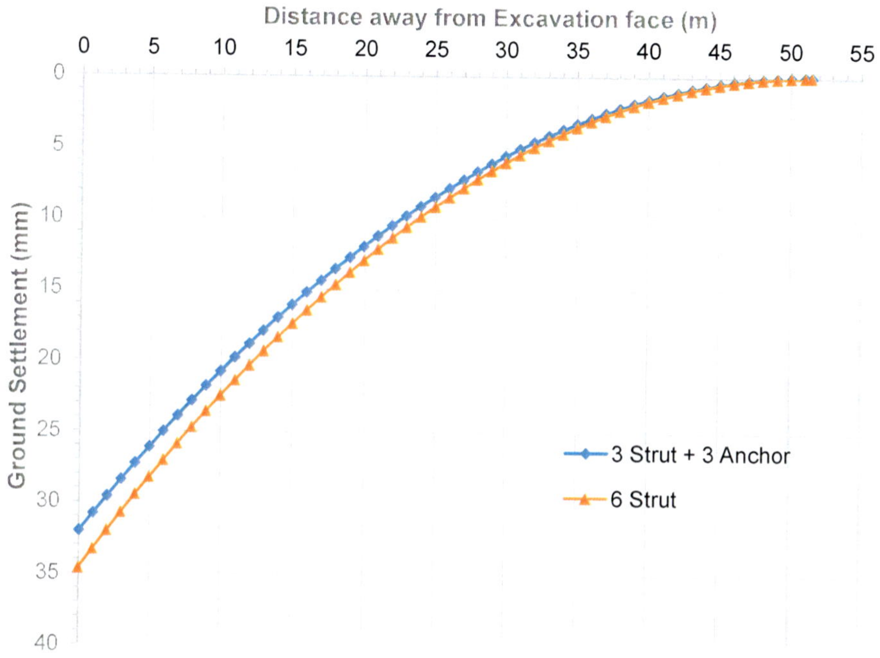

Fig. 6 Ground settlement profile behind the secant pile wall

the ground settlement depends on the lateral movement of the retaining structure. When the deflection of secant piles is high, the maximum ground settlement behind the secant piles will also be high. Figure 6 represents the ground settlement profile behind the secant pile wall.

From Fig. 6, the ground settlement behind the secant pile wall shows that there is an increase in settlement by 8% when struts replace anchors. Similarly, the buildings, which fall within the influence zone of the excavation, will experience differential settlement due to variation in the ground settlement. Due to differential settlement, rotation of the building may occur. Theoretically, the rotation of the buildings can be calculated with the help of the ground settlement profile. There are two buildings which are located at a distance of 3 m and 13 m from the excavation boundary. The details of the building in the zone of influence are mentioned in Table 6.

As the ground settlement follows parabolic distribution, the settlement of the building varies along the length of the building. The rotation of the buildings for

Table 6 Building details near the underground station

Building ID	Dimension	Type of building	Distance from excavation boundary (m)
B1	38 m × 58 m	RCC Framed structure	3
B2	32 m × 34 m	RCC Framed structure	13

Table 7 Summary of building rotation

Building	Building rotation	
	Three struts and three anchor	Six struts
B1	1 in 1960	1 in 1810
B2	1 in 1695	1 in 1565

both the cases is presented in Table 7. The buildings tend to rotate more in a six strut configuration as the distribution of ground settlement is slightly higher than in strut and anchor configuration.

7 Summary

Numerical results from WALLAP shows that the increase in deflection of the secant pile wall is due to a change in the support system from three levels of struts and anchors to six level of struts. The increase in secant pile deflection is about 13% but the deflection pattern looks the same for both the cases. Similarly, the increase in strut forces for the first three levels of struts is about 5%, 10% and 17%, respectively. The increase in the strut force is due to a change in the stiffness of secant piles and the supports. Having the cross-sectional area of the struts very high when compared to anchors, the increase in the deflection and strut forces is because of the restraint provided by the struts. The moment when anchors are replaced by struts, it imparts too much resistance and support to the secant piles. During backfilling sequence, the deflection of secant piles increases when every level of the strut is removed, which in turn increases the forces on the other struts. In a similar fashion, the ground settlement behind the secant piles also increases when the deflection of the secant piles increases. When anchors are replaced by struts, an 8% increase in the maximum ground settlement is observed due to the movement of secant piles. On the other hand, if the buildings are near the influence zone, the rotation of those buildings is high when struts replace anchors. If the struts are placed at the same level where anchors are there, then the response of secant piles may vary and the ground response may be the same for both the cases. However, in the present study, there are certain limitations like stresses induced, potential cracks developed and tensile strains induced in the building during excavation are not accounted in this study.

References

Boscardin MD, Cording EJ (1989) Building response to excavation-induced settlement. J Geotech Eng, ASCE 115(1):1–15

Bowles JE (1990) Foundation analysis and design, 4th edn. McGraw-Hill book company, New York, USA

Burland JB, Wroth CP (1974) Settlement of buildings and associated damage. Proceedings of Conference on settlement of structures. Pentech Press, London, England, pp 611–654

Burland JB, Broms BB, de Mello VFB (1977) Behaviour of foundations and structures. Proceedings of the 9th International Conference on Soil Mechanics and Foundation Engineering, vol 2. Tokyo, Japan, pp 495–546

Clough GW, O'Rourke TD (1990) Construction induced movements of in-situ walls. Specialty conference on design and performance of earth retaining structures, ASCE special publication No. 25:439–470

Timoshenko SP (1955) Strength of Materials, Part I, Elementary Theory and Problems, D. Van Nostrand Company, 3rd Ed.

Overview of Enabling Works for Waterfront Structures—Design and Construction

K. Raja Rajan, D. Nagarajan, and T. Vijayakumar

Abstract Major bridges crossing mighty rivers must be constructed in flowing water. Constructing the sub-structure in water has been a challenging job for the contractor. Special construction techniques with marine fleets are to be adopted for waterfront construction. Enabling works like temporary piling platforms and cofferdams are used for pile and pile cap construction. Apart from this, enabling works like load out jetty to transport the materials from land to water; temporary walkway for access of workman to work location; temporary access bridge for construction vehicles movement; temporary liners for tower crane foundation; Concrete block to act as dead man anchor for barge movements are all required by the contractor for smooth functioning of the site as per construction schedule. Investment in enabling works by the contractor plays a significant role in the profit margin of the project, and of course, with utmost safety. Design of enabling works for waterfront structures involves hydrological data like afflux, bathymetry survey, current force, scour depth, and wave force to be taken cautiously for safe and economic design. The type of foundation and pile/well cap top level corresponding to water levels influence the construction scheme. The usage of geotechnical software like Wallap and Plaxis required for enabling work design to enhance the safety of the structure. Bathymetry and soil condition play a critical role in the design and construction of enabling works. Water Discharge quantity in the river along with water levels like low-water level, high flood level, and seasonal fluctuation of water levels has a great impact on the design of enabling works. Establishing enabling works for waterfront construction near an existing bridge is an additional challenge for the contractor. Along with the design, constructing the enabling works in water requires a special construction methodology and sequence of work. This paper provides insights into the design and construction of enabling works for waterfront construction.

K. Raja Rajan (✉) · D. Nagarajan · T. Vijayakumar
CMPC division, EDRC, Chennai L&T, India
e-mail: k-raja@lntecc.com

D. Nagarajan
e-mail: nagarajan-d@lntecc.com

B. Adimoolam et al. (eds.), *Deep Foundations for Infrastructure Development in India*, Lecture Notes in Civil Engineering 315, https://doi.org/10.1007/978-981-19-8598-0_13

Keywords Afflux · Bathymetry · Cofferdam · Enabling works · Scour depth · Sheet pile · Software · Temporary Platform · Waterfront construction

1 Introduction

To achieve a 5 trillion dollar Indian economy by 2025, India has decided to spend approximately 100 lakh crore on infrastructure projects over the next 5 years which will pave the way to an increase in jobs, ease of living, and improved infrastructure that will attract more investors toward our country. Capex amount shall be shared by the central and the respective state governments by 39% equally, and the rest 22% shall be by shared the private sector. At present, our country's growth is projected as 4.8% in the first half of FY20 and there may be a marginal increase in the second half of FY20 due to various economic disruptions within the country and globally by the pandemic effect. Current disruptions are only temporary, India will soon regain its growth and focus on infrastructure development. Out of 100 lakh crores, around 19% of the amount is planned to be spent on road projects in the upcoming 5 years.

Rivers are the lifeline of our Indian sub-continent. Bridges are always considered as the economy boosters of both connecting places due to increases in cash flows complementing each other economically. Both road and rail bridges connecting various parts of India pave for easy cargo transportation. The present government takes various steps to renew inland waterway transportation. Due to the increase in inland waterways transportation, there is a huge demand for long-span bridges. The engineering firm to which the authors belong has recently successfully completed the construction of the world's longest extra dozed cable bridge at Dhurgam Cheruvu Lake in Hyderabad. Few of the iconic bridges like Mondovi bridge at Goa, Narmada bridge at Bharuch were some of the project to exhibit the construction capabilities of their firm's river bridges construction. Currently, the same firm is carrying out construction works for the Mumbai Trans Harbor link project over the Arabian Sea, the Kachi Darga bridge project over the Ganga river, Second Ishwar Gupta Setu over the Hoogly river to specify a few of the ongoing river bridge projects.

Contractor competency is required for sub-structure construction in river bridges. Marine fleet's control will be of supremacy for speedy construction in water. A special construction method is required for waterfront construction. Many constraints like flash flood, monsoon season, workings in High tide level, etc., make the contractor plan meticulously the construction schedule. Investment in enabling works decides the contractor's profit in the project, so a detailed design of enabling works on waterfront construction is required by the contractor to maximize the profit with the utmost safety of the laborers and structure.

This paper provides an overview of temporary enabling works on waterfront structures from a contractor perspective for a typical bridge project rather than focusing on a single project. Various hydrological and design parameters have to be considered for designing the enabling works along with its construction methods have been discussed in this paper.

2 Hydrological Parameters (Project documents and reports on site-based data for waterfront structures xxxx)

Hydrological parameters are more critical to the design of the structure and for enabling works design. Many parameters like scour depth, afflux, current force, wind force on structure, water level variation, profile of riverbed, nearby existing structures, etc., shall be critical for enabling works design on waterfront construction (Nikonorov et al. xxxx).

Current force: Any part of the temporary structure which is submerged in running water shall be designed to sustain the horizontal pressure due to the water current. The current force is calculated as per IRC 6. Current force is calculated from the deepest riverbed level or deepest scour level to high-water level. The velocity of water current is normally mentioned in the contract document, if not, the same has to be measured at the site. The resultant current force shall be added to the horizontal load and the capacity of the member shall be verified.

Wind force: Wind force is calculated as per IS 875–Part 3. Base wind pressure is selected based on the topography of the project location. The exposed area of temporary works like liners, sheet piles, etc., is calculated for wind pressure and the resultant force shall be added to the horizontal force. If there is variation between high-water level and low-water level, then the current force up to high-water level is to be calculated and wind force is to be calculated up to low-water level. Both the combinations shall be used in the design of enabling structures.

Afflux: Afflux means the rise of water level due to the obstruction of water flow by any temporary or permanent structure. For the permanent pier location, the client/permanent designer would have calculated afflux due to the pier and pile cap. But, as a contractor, for sub-structure construction, any temporary structures like piling platform or cofferdam shall be in place till construction, so afflux effect due to enabling works to be calculated. Afflux calculation is carried out as per IRS code (Bureau of Indian Standards xxxx). For afflux calculation, the entire obstruction area of all temporary structures like cofferdam, temporary walkway, load out jetty, etc., are to be calculated, and then the temporary rise in water level due to enabling works is to be estimated. Based on afflux readings only, the top level of enabling work is finalized so that no water intrusion is possible into the enabling structures work area. Normally, designers tend to fix the top level just above the high-water level, but due to the afflux effect, the increase in water height may lead to the entry of water into the work area. So the top level of enabling works to be finalized is based on high-water level plus the water level increase due to afflux.

Water level variation: Low-water level and high-water level are the two different water levels normally recorded as the hydrological parameter. The difference between these two levels is water level variation. The high water level provided might be the water level recorded during the floods. So, it is the contractor's responsibility to study the water level with respect to seasonal fluctuation. Based on the maximum water level, the top level of enabling works is to be fixed. For example, to fix the temporary walkway, the top level is to be fixed diligently with respect to the water

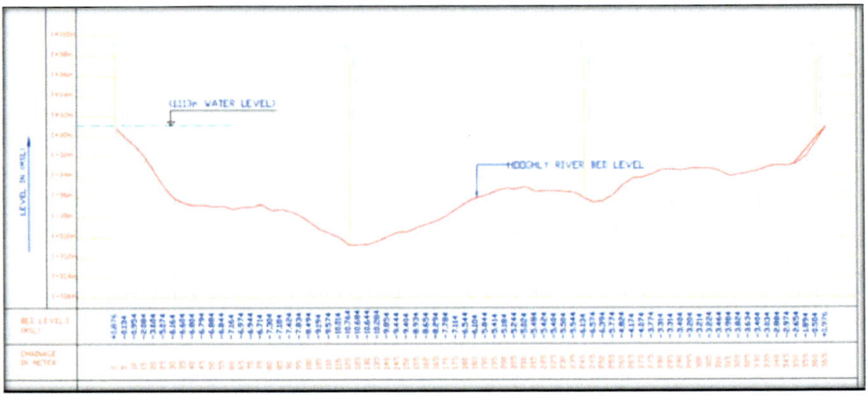

Fig. 1 Typical bathymetry of hoogly riverbed with pier locations

level for all season workings. In some cases, if the contractor wants to work during the off-monsoon season, then the contractor may take the advantage of low-water level, and the risk bounded is very high, since there may be an unprecedented delay or flash floods which is not in the hands of the contractor.

Profile of Riverbed: To know the profile of the riverbed, a bathymetry survey is to be carried out by the contractor. Sometimes the bathymetry shall be provided along with the contract document. Bathymetry is normally carried out at the center of alignment of the proposed bridge and at defined intervals in both upstream and downstream of the river. It is advisable to carry out the bathymetry survey up to the work area of the contractor so that the marine fleet's movement shall be decided based on the draft available. A typical riverbed profile is provided in Fig. 1 by a bathymetry survey which is carried out at the center of alignment.

Riverbed level variation is provided along the center line and pier marking is also provided for better understanding. The graph is plotted between change in ground level and chainage (i.e., from one side of the embankment to the other side). Water level and current force shall also be provided during the time of the bathymetry survey. Below fig shows a planned view of the bathymetry survey carried out at every 10m interval from one embankment to other embankments. Readings are taken every 10 m on either side of the proposed bridge and then the riverbed profile is developed.

At a few project sites, due to high scouring action, the profile of the riverbed changes from season to season. Some depositions from the upstream sides also get settled and change the profile of the riverbed. Contractors have to keep a record of changes in bed profile in order to maneuver the marine fleets and also to establish the enabling works (Fig. 2).

Scour Depth: It is one of the most significant factors which triggers the failure of the foundation of river bridges. Scour depth is defined as the erosion of soil at the riverbed. Scouring increases as the velocity of the water increases (Pizzaro et al 2020). It also depends upon the type of soil, effective diameter of the soil, and design discharge of the river. From IRC 78 (Bureau of Indian Standards xxxx), the below formula is used to estimate the scour depth.

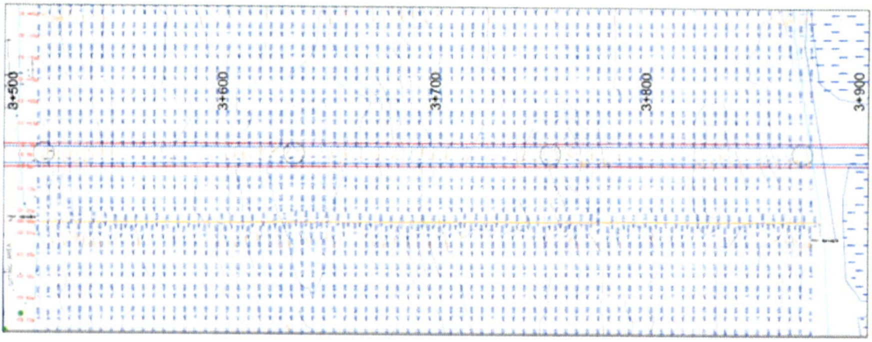

Fig. 2 Typical bathymetry of hoogly river in plan with respect to alignment

$$d_{sm} = 1 \cdot 34\left(\frac{Db^2}{k_{sf}}\right)^{1/3}$$

For enabling work design, the design discharge has been taken only for the last 5 years. Whereas, for the permanent design of the bridge, the design discharge would have been taken for 100 years. Since the enabling works are temporary, the last 5 years discharge value is referred to for estimating the scour depth. Silt factors and other parameters referred from the geotechnical interceptive report. For enabling works, scour depth is calculated from the recorded high-water level and multiplied by a factor of 2 as per IRC 78 code (Bureau of Indian Standards xxxx). The reason for choosing the last 5 years' discharge is to optimize the enabling work quantity and as well as accounting scour depth to some extent. Every time, bathymetry survey cannot be taken to monitor the bed level, so the easy way to have an observation on scour depth is just by checking the sound frequently. Sound readings provide a handful of data to the contractor for the planning of enabling works. Adopting the permanent pile scour depth for enabling work may not be feasible and economical.

3 Cofferdam

Cofferdam is a temporary watertight structure used for the construction of a pile cap. If the bottom of the pile cap is above the high-water level, then possibly cofferdam may be eliminated. Whereas if the pile cap bottom is below water level, then a cofferdam is the only solution to construct a pile cap. Many designers are matching the top of pile cap level to low-water level, so the requirement for cofferdam is increasing among contractors. To have a good architectural view, the pile cap is designed to be underwater or below bed level, whereas if the navigational channel is provided, then the possibility of the marine fleet may hit the pile cap which is below water. Further arrangements are to be provided for the pile cap which is present in the navigational channel.

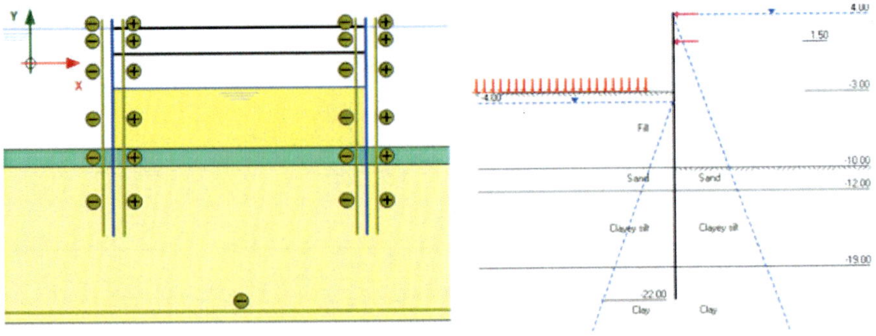

Fig. 3 Cofferdam design by PLAXIS and WALLAP software

Cofferdam is a steel structure with a steel sheet pile or steel liner depending upon the depth of water. Different types of sheet piles like U-piles, Z-pile, combinations of H-beam and Z-pile, double wall sheet pile, etc., are available from various standard sheet pile vendors. If the water depth is more, then the possibilities of steel liners shall be worked out. Structural steel struts are provided within the cofferdam as intermediate supports. Cofferdam design is usually carried out with the two available geotechnical software called PLAXIS 2-D and WALLAP. Based on bending moments, the section and grade of the sheet pile are chosen, and based on strut forces, structural steel members are fabricated and installed, global factor of safety is ensured with sufficient embedment depth. Filling may be considered inside the cofferdam till the pile cap bottom level (Fig. 3).

Cofferdam construction will be in marine mode. A vibro hammer is used to drive the steel sheet pile. The selection of a Vibro hammer depends upon the requirement of centrifugal force and amplitude based on the soil strata. Vibro hammer shall have sufficient capacity to drive the sheet pile till the designed founding level. Barges Dead Weight Tonnage (DWT) shall be calculated considering crane, Vibro hammer, compressor, etc., and suitable size and capacity of the barge to be brought at the site. Movement of barges shall either be by using a tow boat or with the help of a winch tied to a dead man's anchor.

4 Piling Platform

Piling platform is a temporary structure that is erected in the pier location over temporary liners. Temporary liners are positioned carefully between the proposed permanent piles. The friction capacity of driven liners is not to be affected during the permanent piling process. Over the temporary liner, using structural beams and panels, a temporary platform is erected. This platform is used for rig movement to all piling positions. Floating barges are used to stack the permanent liners, reinforcement cage, and Vibro hammer. It is economical and safe to have a crane over a floating barge rather than on a piling platform. If the crane is to be mounted on a piling platform, then accordingly, the platform is to be designed and it may require more operational space. Platform to be designed for base pressure of piling rig and live loads foreseen over the platform. Temporary liners to be designed for the vertical load from the platform adding to the current force, wind force, and scour depth if any shall be accounted. Bracings on either side are to be provided between temporary liners.

Piling rig or A-frame shall be used to bore the pile. If a piling rig is proposed, then the piling rig shall be mounted over the piling platform. Transportation of the piling rig from the floating barge to the piling platform needs marine specialist assistance (Fig. 4).

Many factors like piling rig weight, water level, current flow, inclination of platform for marching piling rig, etc., are to be taken care of while transporting the piling rig. Feeding of materials during the piling process shall be over floating barges. Bored soil shall be stacked over a barge or platform and then disposed of as per contractual requirements. The entire platform shall be removed after the completion of the piling procedure. Floating barges, Vibro hammers, and cranes shall be used for erecting and dismantling the temporary piling platform.

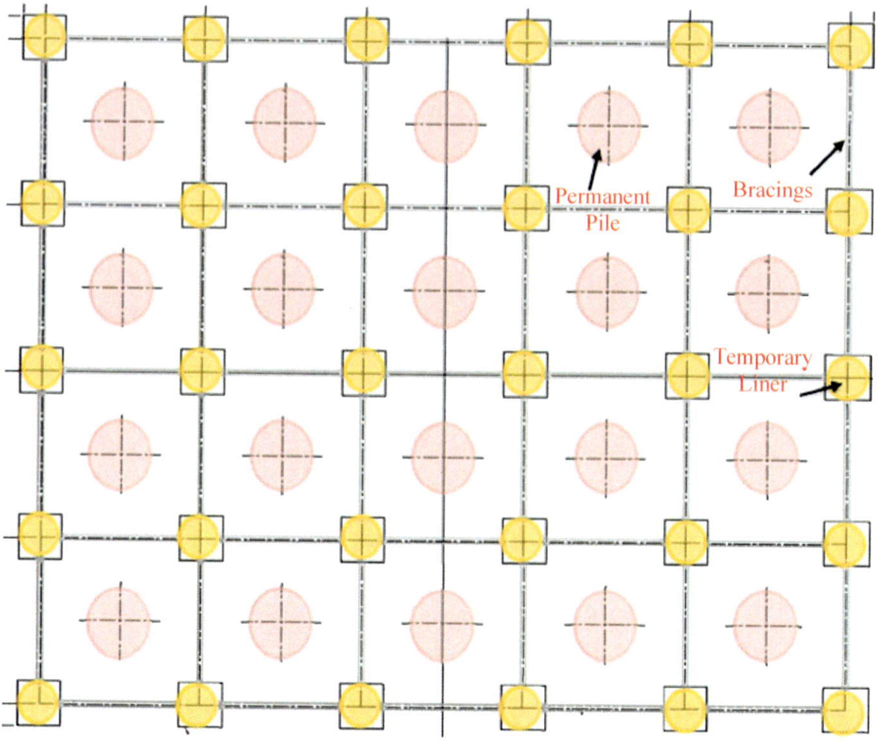

Fig. 4 Plan View of permanent piles and temporary liners for piling platform

5 Temporary Access Bridge

A temporary access bridge is built along the alignment for the entire river crossing stretch. This facilitates the construction equipment movement, labor movement, material movement, etc., from the land portion to the respective pylon/pier position. The temporary access bridge is supported by liners and structural beams with steel or precast concrete panel. The width of the temporary access bridge depends upon the volume of the work to be carried out within the given time schedule. Top level of the temporary access bridge depends upon the high-water level. This temporary access bridge shall be accessible for all monsoon periods. Standard IRC loading is taken for the design of temporary access bridge. This temporary access bridge can be built by moving the piling gantry to drive the liners and service crane to erect the superstructure panel.

6 Load Out Jetty

Load out jetty is a temporary structure used to transfer materials from land to water. Load out jetty comprises temporary liners over which a gantry will be erected to transport the materials. The capacity of the gantry is to be determined based on the forecast of the weight of materials to be shifted till the completion of the project. Load out jetty shall be spanned between land and water so that the gantry can lift the material from the trailer and load it over the barge. Gantry used to travel on rail beam which will be supported by liners at definite spacing. Liners are to be braced to resist the lateral forces.

The Riverbed profile is to be analyzed before setting up the load out jetty. A sufficient draft shall be available for decking the barge in the load out jetty. If sufficient depth is not available, then local dredging is to be carried out to ease the barge movement. Due to seasonal variation, the chances of deposition of sand are very much possible, which has to be monitored frequently. Some fender-type arrangement may be provided to the liners which are located on the waterside, so that direct collision of the barge on the liner can be avoided (Fig. 5).

7 Temporary Walkway

Temporary walkway is used as access for laborers and engineers to reach the work location across the river. Normally, the width of the temporary walkway shall be 1.5 to 2 m which allows access only for man movement and not for any vehicle movement. This is called a lifeline for a contractor till the completion of the project and the same shall be accessible during all monsoon seasons. By installing this temporary walkway, it will largely reduce the barge or small boat movement for

Fig. 5 Typical photograph showing temporary walkway and load out jetty

labor transportation. Adding to the labor movement, small pipelines are laid over this temporary walkway for bentonite and concrete transportation. These pipelines are connected at the embankment and with a pumping motor, bentonite and concrete are pumped to the desired location.

Temporary walkway is designed such that it stands on a single liner at definite spacing connected by a beam or truss over liners. If water depth and current velocity are more, then the single liner may not be feasible, and the contractor left with the option of a double liner or a larger diameter liner. Liner to be designed for lateral forces and to be ensured that there will not be any direct collision by marine fleets. High possibility of the barge hitting the liner is possible, and the contractor is to take care of the same. If any navigational channel is present, then a temporary walkway is to be provided accordingly matching the navigation span.

8 Summary

This technical paper has brought the overview and significance of temporary waterfront structures in bridge construction projects and highlights the hydrological parameters clubbed with the design of enabling works. The construction aspect of temporary waterfront structures has also been briefed and discussed. Design and construction aspects of major enabling works like piling platform, cofferdam, temporary access bridge, temporary walkway, load out jetty which is envisaged in a typical river bridge project has been discussed to bring the contingency in enabling works design.

9 Conclusion

Regarding Temporary Waterfront structures from the contractor's perspective, the following points may be noted:

- Scour depth shall be calculated for enabling works by taking discharge for limited years to economize the temporary structure. Scour depth shall be monitored frequently by maintaining the records of soundings at pre-defined locations.
- Sufficient data like water level, flash floods, and riverbed profile are to be collected from the nearest Port office and the same records are to be maintained till the completion of the project.
- Profile of riverbed to be ensured before planning for temporary waterfront structures. Marine fleet movements feasibility to be carried out with bathymetry.
- Current force, wind force, afflux, scour depth, and other hydrological parameters are to be considered in designing the enabling works.
- Cofferdam, piling platform, Temporary walkway, load out jetty to be designed considering all hydrological parameters and for the foreseen loadings.
- Top level of enabling works is to be decided based on water level variation and afflux calculation. Care is to be taken such that water shall not enter the work area.
- The usage of the temporary walkway and temporary access bridge for a project is based upon the location of the project and the time schedule of the project.
- Movement of marine fleets is to be carried out with marine specialist, and direct collision is to be avoided with temporary waterfront structures.

Acknowledgements The authors would like to convey their thanks to the management for allowing them to present this study; the authors also convey regards to colleagues and the site team for their valuable contribution.

References

Bureau of Indian Standards IS 2911-Part I-Sec2, IRC 6, IRC 78, IS 10084-Part I and IRS code
Nikonorov A, Pavlov S, Terleev V et.al. "Use of Enclosing and Temporary special structures under the reconstruction of hydraulic facilities in Saint- Petersburg". From www.sciencedirect.com, Saint-Petersburg, Russia
Pizzaro A, Manfreda S, Tubaldi E (2020) "The science behind scour at Bridge foundations – A Review". From Water, www.mdpi.com/journal/water; published on 30th January 2020
Project documents and reports on site-based data for waterfront structures

Realistic Estimation of Water Table Depth for Design Optimisation of Bored Tunnel and Cut and Cover Structures for Underground Metro

Chiranjib Sarkar, Sibapriya Mukherjee, and NKumar Pitchumani

Abstract With the rapid expansion of urban transportation systems, tunnels and cut & cover structures are considered as the only solution in improving the urban space congestion problem in mass rapid transit system. Therefore, it is necessary to accurately design underground structures with realistic assumptions and considerations of design parameters like site geotechnical data, water table, surcharge load, etc. Over-conservative approach provides not only uneconomical design but also frequently results in overdesign of the structures. Water table depth has a significant role in some design aspects of underground structures like floatation check, lateral and uplift pressure on the buried structure, etc. In the current practice of tunnel and cut & cover structure design, water table is assumed to coincide with the ground level. Most of the Design Basis Reports & Outline Design Specifications directly mention consideration of water table at ground level for floatation check and load calculations, etc. However, the actual scenario is different in most of the cities, especially in northern, central, western and eastern regions of India except the coastal cities. In the present study, an attempt has been made to carry out a parametric study on the effect of water table depth in the design of a typical 6.3 m outer diameter circular tunnel with a 6 m backfill. In this study, it is also attempted to establish an analytical method of calculating the most realistic consideration of water table depth instead of the present hypothetical assumption of considering water table to coincide with the ground level. The findings of the current study may be helpful to the researchers and practising engineers in the design of tunnel and cut & cover structure for subways and metros.

Keywords Underground structure · Water table · Floatation

C. Sarkar (✉)
AECOM India (P) Ltd, Kolkata, India
e-mail: chiranjib.sarkar@aecom.com

S. Mukherjee
Jadavpur University, Kolkata, India

N. Pitchumani
AECOM India (P) Ltd, Chennai, India
e-mail: NKumar.Pitchumani@aecom.com

B. Adimoolam et al. (eds.), *Deep Foundations for Infrastructure Development in India*, Lecture Notes in Civil Engineering 315, https://doi.org/10.1007/978-981-19-8598-0_14

1 Introduction

Underground infrastructure is being extensively used to tackle urban space congestion arising from land insufficiency. Metro railway is considered as a lifeline for development of any city. A metro rail project should be planned and designed in such way that it can provide a socially, economically and environmentally sustainable transport network. Hence, the design of tunnels and underground structures should be done with realistic assumption and actual site-specific consideration of design parameters like site geotechnical data, water table, surcharge load, etc. It is a well-known fact that over-conservative approaches provide uneconomical and unrealistic situations. Water table depth has an important role in some design aspects of tunnel and underground structure like liquefaction analysis, floatation check, applicability of lateral and uplift pressure on the buried structure, etc. In the existing practice of design of a tunnel and cut & cover structure, water table is assumed to coincide with the ground level, whereas actual level of water table is totally different in most of the cases. Like soil profile, water table varies from location to location and project to project. Each project and location has its own uniqueness in water table depth which depends on many factors and parameters. Therefore, water table or hydrostatic pressure consideration in design should be project specific. The existing common and typical concept of considering water table at ground level is hypothetical, over-conservative and misleading. There is high time now to rethink or to start comprehending realistic consideration of water table in design of tunnel and underground structures. The present paper suggests that water table depth should be considered as project-specific instead of generalising for all projects. Water table depth has a significant impact on the following design aspects of tunnel and underground structures:

(a) Lateral and uplift pressure,
(b) Floatation,
(c) Overburden pressure and ground deformation and
(d) Liquefaction analysis.

A. Lateral and uplift pressure

Hydrostatic loads (lateral pressure and vertical uplift) applied on tunnel and underground structure are directly proportional to the depth of water table. Intensity of hydrostatic load/water pressure (lateral and uplift) varies based on water table depth below the ground. It increases when the water table depth below ground reduces, whereas it decreases when the depth of water table increases. Hence, the selection of proper water table depth from the ground is an important part of the design of tunnel and underground structures. A schematic view of application of lateral water pressure and vertical uplift pressure on cut & cover box tunnel structure is shown in Fig. 1. Lateral water pressure acting on cut & cover box tunnel wall for different water table depths (water table at ground, water table at 2 m below ground and water table at 4 m below ground with the RL of the ground level being 118.5 m) are shown in Fig. 2

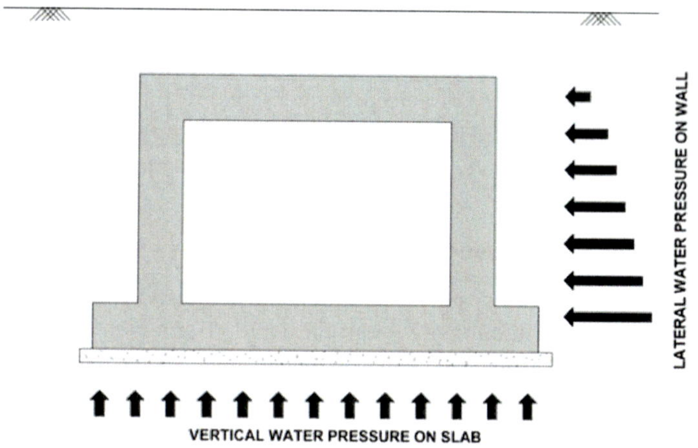

Fig. 1 Schematic view of lateral water pressure and vertical uplift pressure acting on cut & cover box tunnel structure

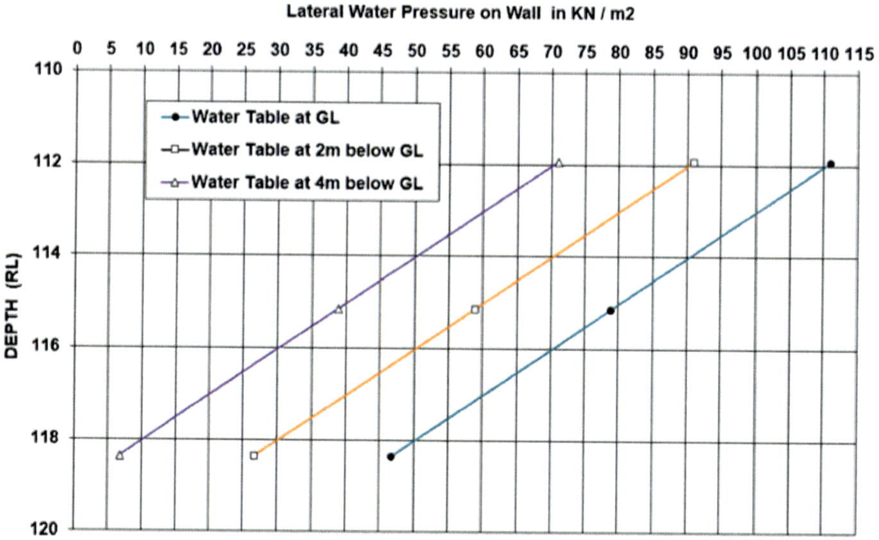

Fig. 2 Comparison of lateral water pressure acting on cut & cover box tunnel wall for different water table depths (water table at ground, water table at 2 m below ground and water table at 4m below ground)

B. Floatation

The present trend for flotation check is to assume water table coincidence with ground level (Report of the Ground Water Resource Estimation Committee (GEC-2015) 2017). This seems to be hypothetical. In most of the cases, the actual water table is totally different from the present consideration and it varies from location to location and project to project. As per the current concept for floatation check, the design assumes that the entire structure is submerged under water and a buoyant force is acting as uplift force on the bottom of the structures, but the actual scenario is much different in the case of a deeper water table. It is a fact that the water table never coincides with ground level (in the case of higher water table) even in flood or any adverse situation. Hence, it is necessary to adopt an appropriate depth of water table for flotation check.

C. Overburden pressure and ground deformation

Overburden pressure and ground deformation have a direct relationship with water table depth. With the variation of water table depth, overburden pressure and ground deformation also fluctuate. Hence, selection of proper water table depth from the ground is required for the calculating of actual overburden pressure and ground deformation.

D. Liquefaction analysis

Liquefaction Potential is directly related with effective vertical stress which is also dependent on water table depth. As per the current practice of liquefaction analysis, it is assumed that the water table is coincident with ground level. This is not the correct approach. Water table should be location and project specific instead of taking it at ground level. Hence, it is essential to adopt a suitable depth of water table for liquefaction analysis

In the ongoing practice of design of tunnel and cut & cover structures, water table is assumed to coincide with the ground level. Most of the Design Basis Reports (DBR) & Outline Design Specification (ODS) for metro projects have directly stated to consider water table at ground level for liquefaction depth calculation and floatation check (Model design basis report (DBR) for bored tunnel sections of metro system in India. Ministry of Railways (Railway Board), Government of India, March 2017; Corporation 2019). Furthermore, design load and load conditions can be referred from the guideline mentioned in 'Model design basis report (DBR) for bored tunnel sections of metro system in India'. As per guideline, design loads and load cases would be calculated by considering water table at ground surface (Model design basis report (DBR) for bored tunnel sections of metro system in India. Ministry of Railways (Railway Board), Government of India 2017). But water table at tunnel stretch varies from location to location and project to project. Therefore, the actual scenario of most of the cities in India (except the coastal cities) are different from what is mentioned in the guideline for design of tunnel and underground structure. In the present paper, an attempt has been made to find out the effect of water table depth in design of a typical cut & cover box tunnel and a circular tunnel with 6m backfill. The present study also encompasses an analytical method for calculating the most realistic consideration of water table depth instead of the present hypothetical assumption of water table at the ground level.

2 Model of the study

2.1 *Project-Specific Hydrological Feature Study*

Two projects from two different cities have been considered for this study and analysis. The first one is a typical circular tunnel (6.3 m outer diameter) section of the Delhi Metro Dwarka–Najafgarh corridor. Actual groundwater table has been observed between 18.0 and 19.0 m below ground level during soil investigation. A seasonal fluctuation in water table, of an average of 2 m, has been found for Delhi region (Ground Water Year Book NCT Delhi xxxx). Depth of water table and groundwater contours of the National Capital Territory, Delhi (CGWB, 1995), are shown in Fig. 3.

There are two main aquifers within the route; the alluvium and the rock. Groundwater flow within the alluvium is controlled by inter-granular flow and it is a variable depending upon the spatial form of the historical channel deposits, while flow within the rock is controlled by fissures. Recharge of the aquifers occurs from the Yamuna River catchment via the alluvium, and from the Aravalli Range via the basement rock. The highest flood level (HFL) at the Yamuna River is recorded at 209.86 m, whereas the lowest bed level (LBL) or lowest water level (LWL) is 206 m. Hence, the depth of water at the highest flood level is near about 4 m (3.86 m).

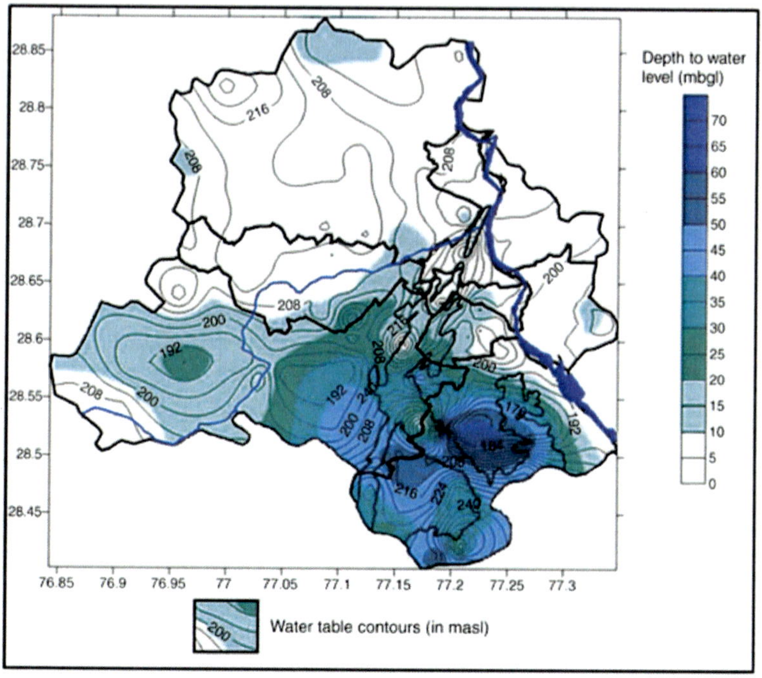

Fig. 3 Depth of water table and ground water contours of National Capital Territory, Delhi

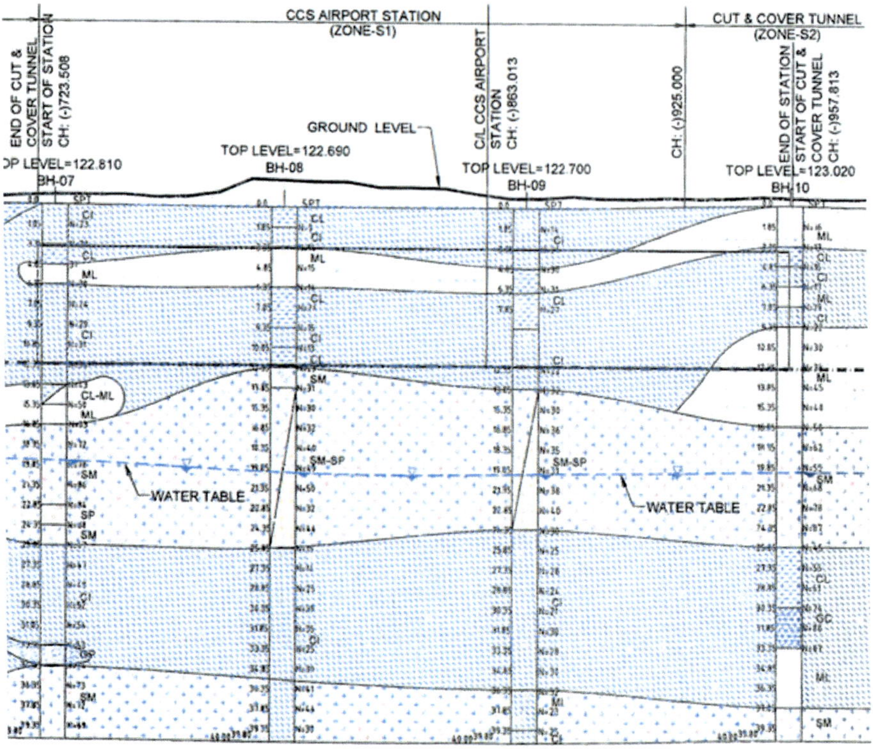

Fig. 4 Long section of geotechnical profile with water table of 18m near CCS Airport station

The second project is a typical cut & cover tunnel (9.6 m wide x 5.8 m high) stretch in the north–south underground corridor of Lucknow metro. Actual groundwater table has been observed near about 18.0m below ground level during soil investigation. Long section of geotechnical profile with water table of 18.0m near CCS Airport station area is shown in Fig. 4.

More than 80% of the land area in Lucknow City is situated on the Central Ganga alluvial plain and stretches on both sides of the Gomti River. The highest flood level (HFL) at the Gomti River is recorded at 111.60 m (in 1960), whereas the lowest bed level (LBL) or the lowest water level (LWL) is 107.4 m (Hydrological Study for Gomti River Front Development 2013). Hence, the depth of water at the highest flood level is near about 4.2 m. Groundwater level zones of Lucknow city area (Ground Water Department, Uttar Pradesh, 1995) are shown in Fig. 5.

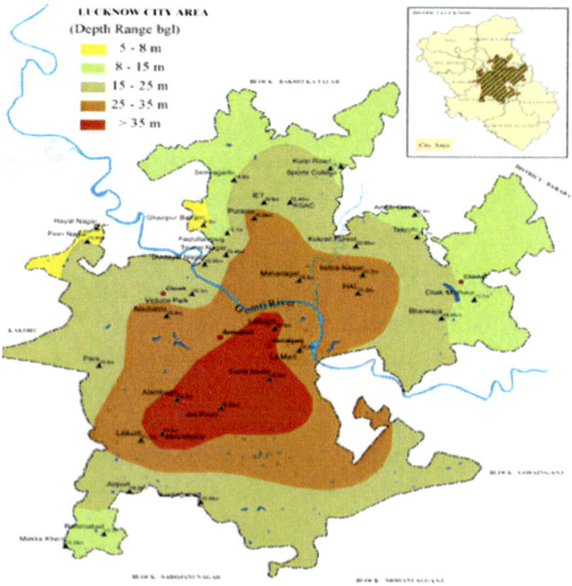

Fig. 5 Ground water level zones of Lucknow city area

2.2 *Parametric Study on the Effect of Water Table*

An attempt has been made to carry out a parametric study on the effects of water table depth in design of a typical 6.3 m outer diameter circular tunnel with a 6 m backfill. A two-dimensional numerical analysis (using finite element method) with 0.5 X 0.5 m mesh size has been developed with geotechnical software MIDAS GTX NX with consideration of water table at tunnel top level and tunnel centre level.

3 Groundwater Estimation Model Study

The assessment of groundwater comprises of dynamic groundwater resources and in-storage groundwater resources.

3.1 *Assessment of Dynamic Groundwater Resources*

Ground Water Estimation Committee-2015 proposed a methodology for the estimation of dynamic ground water resources based on the principle of water balance, i.e. Inflow – Outflow = Change in Storage (of an aquifer). This principle can be further elaborated by Eq. (1) as follows:

$$\Delta S = R_{RF} + R_{STR} + R_{C} + R_{SWI} + R_{GWI} + R_{TP} + R_{WCS} \pm VF \pm LF - GE - T - E - B \quad (1)$$

where,

ΔS–Change is storage of an aquifer
R_{RF}–Recharge from rainfall
R_{STR}–Recharge from stream channels
R_c–Recharge from canals
R_{SWI}–Recharge from surface water irrigation
R_{GWI}–Recharge from groundwater irrigation
R_{TP}–Recharge from Tanks and Ponds
R_{WCS}–Recharge from water conservation structures
VF–Vertical inter-aquifer flow
LF–Lateral flow along the aquifer system
GE–Groundwater Extraction
T–Transpiration
E–Evaporation
B–Base flow

It is a fact that detailed groundwater budgeting of some assessment units is not possible due to the absence of a proper database. Hence, the estimation is carried out by using lumped parameter estimation approach.

3.2 *Assessment of In-Storage Groundwater Resources*

Assessment of in-storage groundwater resources or static groundwater resources can be made after defining the aquifer thickness and specific yield of the aquifer material. Ground Water Estimation Committee-2015 proposed a methodology for the estimation of in-storage groundwater resources which can be further elaborated by Eq. (2) as follows:

$$SGWR = A \text{ x } (Z_2 - Z_1) \text{ x } S_Y \tag{2}$$

SGWR = Static or in-storage Groundwater Resources
A = Area of the Assessment Unit
Z_2 = Bottom of Unconfined Aquifer
Z_1 = Pre-monsoon water level
S_Y = Specific Yield in the in-storage Zone

The sum of annual usable groundwater resources and the in-storage groundwater resources is the total groundwater availability of an aquifer.

4 Results and Discussions

The outcomes of the present study are as follows:

4.1 Effects of Different Water Table Depths on Design of Tunnel at the Same Depths with Same Ground Condition

To address the effect of different water table depths on design of tunnel at the same depth with same ground condition, a 2D finite element analysis has been performed on a typical 6.3m outer diameter circular tunnel with 6m backfill with considerations of water table at tunnel top level and tunnel centre level. Bending moment of typical 6.3m diameter circular tunnel at same depth with same ground condition and water table at tunnel top level is shown in Fig. 6 and with the water table at tunnel centre level is shown in Fig. 7.

From the results of the above-shown figures, it is clear that the bending moment decreases with the increase of depth to water table below ground, even when tunnel depth and ground condition remain the same. As the behaviour of structure changes with the changes of water table depth, therefore, water table should be used as

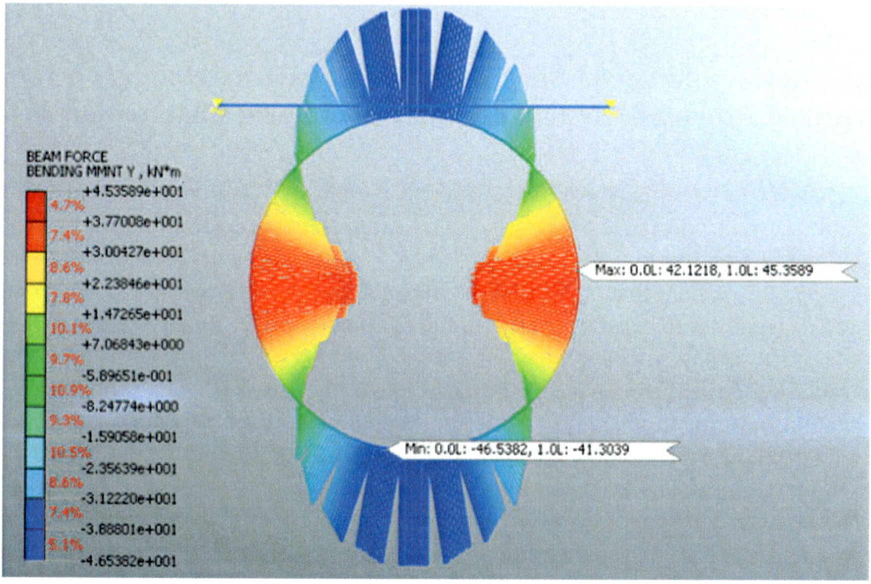

Fig. 6 Bending moment of typical 6.3 m diameter circular tunnel at same depth with same ground condition and water table at tunnel top level

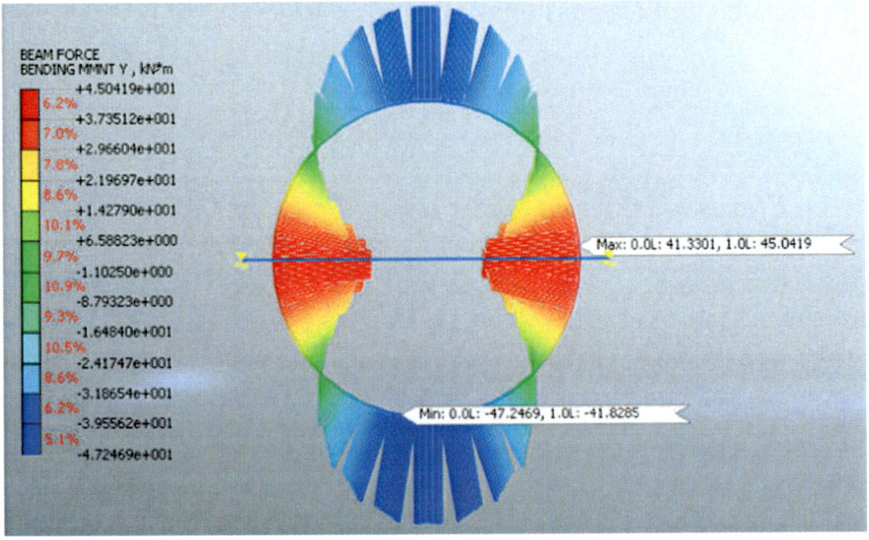

Fig. 7 Bending moment of typical 6.3 m diameter circular tunnel at same depth with same ground condition and water table at tunnel centre level

location- or project-specific instead of the present practice of considering water table at ground.

5 Review of Project-Specific Hydrological Model and Proposed Methodology of Groundwater Estimation

From project-specific actual hydrological data, it can be observed that actual water table is at much deeper levels than hypothetical consideration of water table at ground level as the present concept.

Due to seasonal fluctuation (pre-monsoon and post-monsoon cases), an average of 2m depth would be additionally required to include with actual groundwater table measured during soil investigation at project site. Generally, design life of urban transportation structures is mostly in the range of 100–120 years. Therefore, the possibility of the highest floods and their depth may also be assessed statistically. The depth of water at the highest flood level (HFL−LBL/LWL) would be included with groundwater table.

Based on the principle of water balance, i.e. Inflow − Outflow = Change in Storage, and review of groundwater estimation model, a reasonable water table for tunnel design can be proposed as the sum of actual groundwater table measured during soil investigation at project site, 2m depth for seasonal fluctuation and depth of water at the highest flood level (HFL−LBL/LWL). This principle can be further elaborated by Eq. (3) as follows:

$$\text{Water table depth for tunnel design} = \text{Actual ground water table (measured during soil investigation at project site)} + 2\text{m (for seasonal fluctuation)} + \text{Depth of water at highest flood level} \quad (3)$$

In the case of liquefiable soil, maximum value between the liquefaction depth and the highest flood level depth would be used in place of the highest flood level depth.

Equation (3) can be further illustrated with a working example for a location, where the actual groundwater table is observed at RL (+) 80m (20m below the ground level) and the depth of water at the highest flood level is found as 4m. Thus, water table depth for design will be calculated as (80+2+4) = 86m, i.e. (20−2−4) = 14 m below the ground level.

6 Conclusions

The present study tried to establish an analytical method for proposing a realistic depth of water table instead of the present hypothetical assumption of considering water table at ground level. The conclusions drawn from the current study and analysis are as follows:

(i) Water table should be considered as location- or project-specific instead of the present practice of considering at ground.
(ii) Project-specific water table depth (refer to Eq. 3) is defined as a sum of actual groundwater table at project site, 2 m depth for seasonal fluctuation and depth of water at the highest flood level. This proposed project-specific reasonable water table will have a significant impact in the optimisation of tunnel design in most of the cities, especially in northern, central, western and eastern regions of India except the coastal cities.
(iii) Liquefaction analysis and floatation check should be done with project-specific water table depth (refer to Eq. 3).
(iv) During design load and load conditions calculation, lateral water pressure, vertical uplift pressure, overburden pressure, etc., should be calculated with project-specific water table depth (refer to Eq. 3).
(v) For design of any temporary structure or any temporary and construction stage analysis, actual groundwater table at project site with an additional 2m depth for seasonal fluctuation should be used.
(vi) Moreover, an intensive investigation of project/location-specific hydrological data and upgradation of the current conventional concept is further required in design of structures other than underground like entry, exit, ancillary building, above-ground structure, etc.

References

Report of the Ground Water Resource Estimation Committee (GEC-2015) (2017) Ministry of water resources, River development & ganga rejuvenation, Government of India

Model design basis report (DBR) for bored tunnel sections of metro system in India (2017) Ministry of Railways (Railway Board), Government of India

Ground Water Year Book NCT Delhi 2015-16. Ministry of water resources, river development & ganga rejuvenation, Government of India.

Delhi Metro Railway Corporation (2019) Technical specification: DMRC out line design specification for phase-IV

Hydrological Study for Gomti River Front Development (2013) Dept. of Civil Eng., Indian Institute of Technology Roorkee

Research, Experimental and Numerical Methods in Deep Foundations and Deep Excavation Technologies

Comparison of Ground Movement Near Buildings Due to Underground Station Excavation With Analytical and Numerical Methods

A. Srinivas and C. Anburaj

Abstract In urban areas, construction of underground stations and basements is always a challenging task due to the presence of dense sensitive buildings in the vicinity. The main objective of this paper is to predict the ground movements associated with construction phase of launching shaft excavation in Bangalore Metro project analytically. Also, the comparison of these predicted values with numerical values, and subsequently with actual measurements is presented. These values will help to assess the potential damage, both architecturally and structurally, to the existing building in the influence zone due to ground movements. The principles given by (Bowles JE (1990) Foundation analysis and design, 4th Edn. McGraw-Hill book company, New York, USA. Foundation Analysis and design. 4th Ed., McGraw-Hill book company, New York, USA.) and (Clough and O'Rourke, Specialty conference on Design and Performance of Earth Retaining structures, ASCE Special publication, No. 25:439–470, Clough GW, O'Rourke TD (1990) Construction induced movements of in-situ walls. In: Specialty conference on Design and Performance of Earth Retaining structures, vol. 25. ASCE Special publication, pp. 439-470) depending on the type of soil shall be used to compute ground movements of buildings present in the influence zone. Based on several case histories, (Clough and O'Rourke, Specialty conference on Design and Performance of Earth Retaining structures, ASCE Special publication, No. 25:439–470, Clough GW, O'Rourke TD (1990) Construction induced movements of in-situ walls. In: Specialty conference on Design and Performance of Earth Retaining structures, vol. 25. ASCE Special publication, pp. 439-470) suggested that the settlement profile is triangular for an excavation in sandy soil or stiff clay. The maximum ground surface settlement will occur just behind the wall. The influence zone of the corresponding settlement will extend about twice to thrice of the influence depth (H_e) for sandy soil and stiff to very hard clays, respectively. This paper compares ground movement of building predicted by above analytical method and numerical analysis carried out using PLAXIS 2D

A. Srinivas (✉) · C. Anburaj
L&T Construction, Chennai, Tamilnadu, India
e-mail: a-srinivas@lntecc.com

C. Anburaj
e-mail: anburaj-c@lntecc.com

B. Adimoolam et al. (eds.), *Deep Foundations for Infrastructure Development in India*, Lecture Notes in Civil Engineering 315, https://doi.org/10.1007/978-981-19-8598-0_15

finite element software. The buildings that are in the influence zone of excavation are considered. On comparing the results obtained from numerical, analytical and actual settlement values, it is observed that wall deflection from numerical values is about 33% higher than actual value. In addition, the maximum ground settlement obtained from numerical analysis is comparable with the settlement obtained from analytical approach. However, the observed settlements at the building locations are significantly smaller than predicted. This might be due to support provided by the secant pile walls in the opposite side as the stress around the retaining wall will be in a three-dimensional direction. Also, the deep ground water table during the excavation could have helped in reducing the wall movement and the ground settlement.

Keywords Ground movement · Influence zone · Instrumentation · PLAXIS 2D

1 Introduction

For the construction of underground metro stations, there is a major concern about the influence of the resulting ground movements on the adjacent buildings, structures, pavements or paved roads, utilities etc. In general, during excavation, there is a change in state of stress and the orientation of principal planes. This change in state of stress often leads to induce vertical and lateral movements on the ground. As a result, the structures which are close to the excavation tends to rotate, distort, deform and even lead to damage also. Therefore, for the urban constructional works, the magnitude of the ground movements and the building movements should be within the tolerance limits. Many studies have been made about the building response due to excavation and tunneling. The vertical and horizontal ground movements on the adjacent buildings and structures can be estimated based on Boscardin and Cording (1989) and the limiting tensile strain induced on the adjacent buildings is calculated as suggested by Burland and Wroth (1974). Secant piles are used as the temporary retaining structure with internal horizontal struts for the construction of a shaft to lower or launch the tunnel boring machine (TBM). This paper presents the comparison of ground movement due to construction of Launching shaft in Bangalore Metro project from available analytical methods with numerical results using PLAXIS 2D. For the evaluation of ground movements near excavation, the buildings in the proximity of influence zone are only considered. The analytical and numerical results are further compared with the values recorded at site with the help of instrumentation installed in the adjacent buildings close to the excavation and in the retention systems of the launching shaft.

2 Description of Site

The launching shaft is located in the South of the Shivajinagar Station, which is one of the underground stations in Reach 6 of Bangalore Metro Phase II. The launching shaft is aligned to the south of Gottigere–Nagawara line. Secant piles are proposed and used as temporary earth retaining structures for the construction of the launching shaft. The construction of launching shaft is done by bottom-up method of construction. Struts are used as a support system to hold the secant pile walls. The depth of excavation is about 20 meters and that of water table is about 5 m below the existing ground level. Two buildings are very close to the excavation of launching shaft, which are named as B1 and B2. The strut layout plan of the launching shaft with the adjacent buildings is shown in Fig. 1.

2.1 *Details of the Buildings*

The buildings B1 and B2, which are close to the launching shaft is located at 3m and 13m respectively from the edge of the secant piles. As seen from Fig. 2, both the buildings B1 and B2 are skewed to the edge of the excavation boundary. The general details of the building such as dimensions, height, number of floors, and foundation type are mentioned in Table 1. The details of the building are obtained from the building condition survey (BCS) reports.

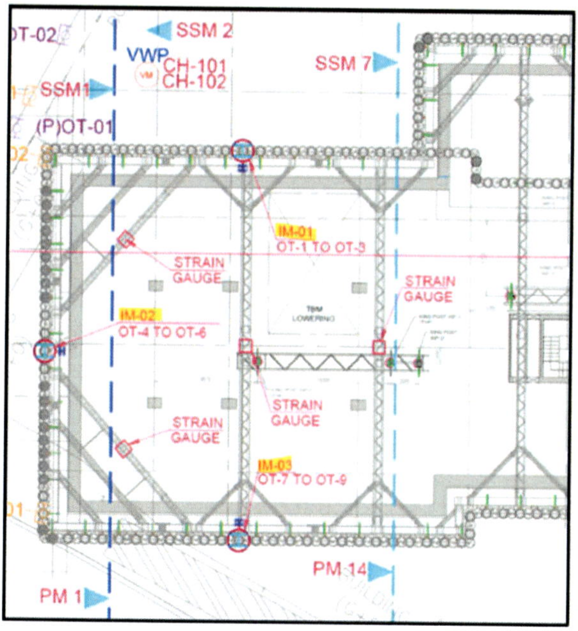

Fig. 1 Plan of launching shaft with instruments

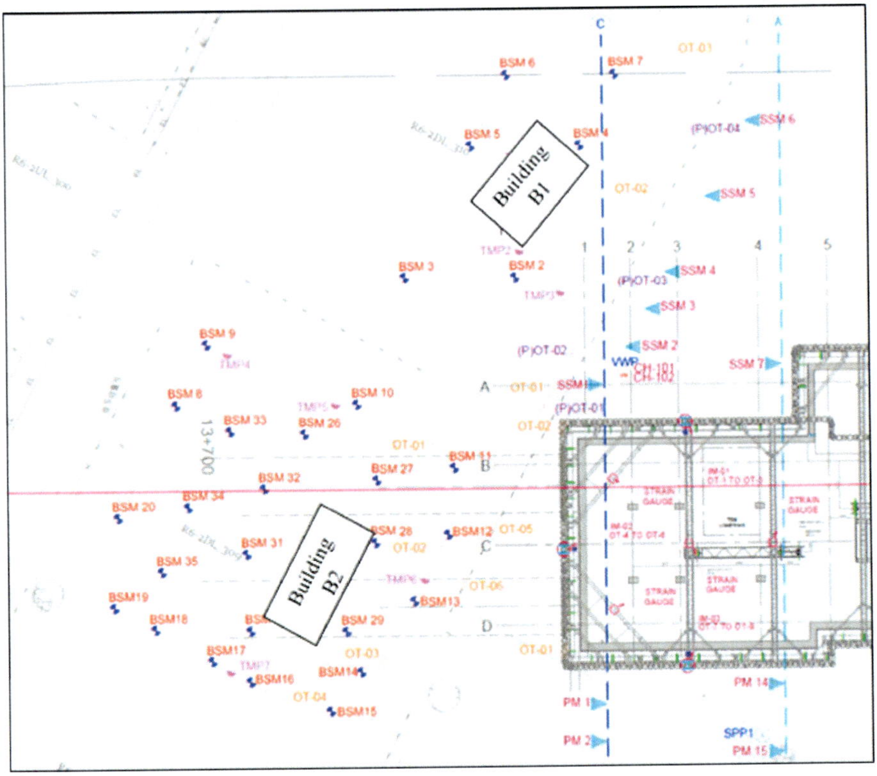

Fig. 2 Plan and orientation of adjacent buildings and the instrumentation

Table 1 Building details near to launching shaft

Building ID	Dimension	Number of stories	Type of building	Distance from excavation boundary	Foundation system
B1	38 m × 58 m	3B+G+3	RCC Framed structure	3 m	Isolated shallow foundation
B2	32 m × 34 m	B+G+5	RCC Framed structure	13 m	Isolated shallow foundation

3 Ground Settlement Based on Analytical Approach

Ground movements can be computed by the principles given by Bowles (1990) and Clough & O'Rourke (1990) depending on the type of soil. Based on several case histories, Clough and O'Rourke (1990) suggested that the settlement profile is triangular for an excavation in sandy soil or stiff clay. The maximum ground surface

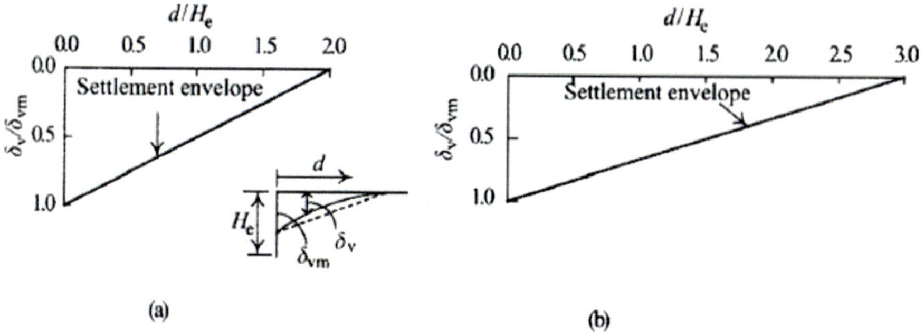

Fig. 3 Dimensionless settlement envelope adjacent to excavation **a** sandy soil **b** stiff clays (Clough and O'Rourke, 1990)

settlement will occur just behind the wall. The non-dimensional profiles given in Fig. 3 shows that the corresponding settlement extends to about 2 and $3H_e$ for sandy soil and stiff to very hard clays, respectively where the influence depth for clayey soil is the depth of excavation plus the width of excavation. However, it shall be limited to 4 times the depth of excavation in the case of very soft soil for a large depth.

As the excavation progresses, the lateral pressure imposed by the ground behind the wall would induce wall deflections into the excavation. This would result in vertical and lateral displacements of the ground adjacent to the retaining wall. In principle, the magnitude and extent of this ground movement is a function of the retention system type, the adopted construction methodology and the properties of the soil and/or rock materials. The depth of influence (H_e) is considered depending on the depth of the secant pile wall and depth of the excavation. The geology of Bangalore is predominantly mixed soil condition and mainly comprises mixtures of silty sand and clays with low to high plasticity and compressibility. The maximum deflection on the ground adjacent to secant pile due to excavation at launching shaft can be estimated with the deflection profile of secant pile. As the strata is generally mixed, the typical settlement profile just behind the secant pile wall is shown in Fig. 4.

The ground settlement curve is taken as "second degree exponential curve" as suggested by Bowles (1986) where the maximum ground settlement occurs just behind the wall. Bowles (1986) suggested a procedure to estimate the excavation induced ground surface settlements using the following relations,

$$\delta_v = \delta_{vm}\left(\frac{l_x}{D}\right)^2 \tag{1}$$

where δ_v = settlement at a distance of $D - l_x$; δ_{vm} = maximum ground surface settlement; l_x = Distance from a point at a distance D from the wall and D is the influence range of ground surface settlement.

The maximum ground surface settlement δ_{vm}, is estimated from the following equation,

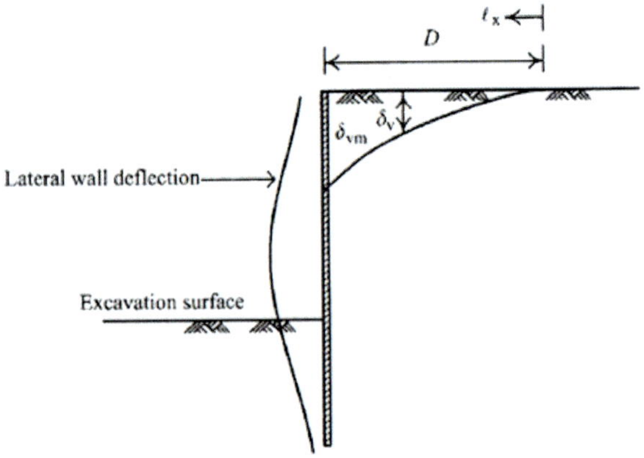

Fig. 4 Estimation of ground settlement (Bowles, 1986)

$$\delta_{vm} = \frac{2A}{D} \tag{2}$$

where A is the area of the lateral wall deflection. The lateral deflection of the secant pile wall is obtained using finite element software, PLAXIS 2D. For excavations which are close to the existing building and structures, differential settlement and angular distortion are the main factors causing damage to structures when the ground settles. The results of the differential settlement and angular distortion for buildings B1 and B2 based on analytical approach are presented in the subsequent section.

4 Numerical Modelling

Numerical analysis constantly proves to be an efficient tool in analyzing complex problems. Analyses are performed considering all the aspects such as construction methodology, foundations of the adjacent buildings and secant pile walls of the launching shaft. The general assumptions involved in any deep excavation problems holds good in this case also. The entire analysis is carried out using PLAXIS 2D with the available data. Soil layers are modelled as linearly elastic perfectly plastic model i.e., Mohr-Coulomb constitutive model to simulate the behavior. The secant piles and the adjacent buildings B1 and B2 are modelled as plate elements. The properties used for the soil layers are mentioned in Table 2. The material properties used for plate elements are mentioned in Table 3.

4.1 Analysis using PLAXIS 2D

As observed from Fig. 1, it is clearly evident that buildings B1 and B2 are skewed to the excavation of launching shaft. Therefore, to simulate behaviour of foundation of

Table 2 Soil parameters used in the PLAXIS 2D

Depth (m)	Strata Type	γ (kN/m³)	E' (MPa)	c' (kPa)	ϕ'
0−2.5	Filled up	18	10	–	28°
2.5−7.5	Clay & silt of low plasticity	18.5	11+3.8z	26	24°
7.5−16.5		18.5	30+3.8z	46	25°
16.5−21	SM	18.5	64.2+3.8z	–	38°
21−26	SC	18.5	81.3+3.8z	15	23°
>26	Soft rock	18.5	136	–	40°

Table 3 Material properties for secant pile wall and buildings used in PLAXIS 2D

Material type	Young modulus (kPa)	Cross sectional area A (m²)	Spacing (m)	Moment of inertia I (m⁴)	EA (kN/m)	EI (kNm²/m)
Secant pile	2.74E6	0.5026	1.35	0.0201	10.2E6	407.9E3
B1	2.74E6	0.6	–	0.018	16.4E6	493.0E3
B2	2.74E6	0.6	–	0.018	16.4E6	493.0E3

the buildings due to response of secant pile is a three-dimensional problem. However, in order to reduce the complexity, it is assumed that the building B1 is aligned perpendicular to the direction of the TBM launching. By doing so, it is easier to model both the building and secant piles in 2D itself. As the excavation is symmetrical, only half of the excavation is modelled in PLAXIS 2D to estimate the ground movements behind the secant pile wall. The cross sections considered for the PLAXIS 2D analysis is shown in Fig. 5.

The entire excavation at Launching shaft location is simulated in the PLAXIS 2D based on stage wise construction sequence. The surcharge of the building and construction surcharge are also modelled before the commencement of excavation at launching shaft. Typical PLAXIS 2D model with soil, secant piles and building foundation with loads are shown in Figs. 6, 7.

In PLAXIS 2D, analysis is carried out up to final excavation of the launching shaft only. The backfilling sequence which is usually being done for any underground metro construction is not considered. This is due to the fact that the maximum deflection in the secant pile is encountered during excavation activity itself. The results of ground settlement behind the secant pile wall from PLAXIS 2D are presented in the next section.

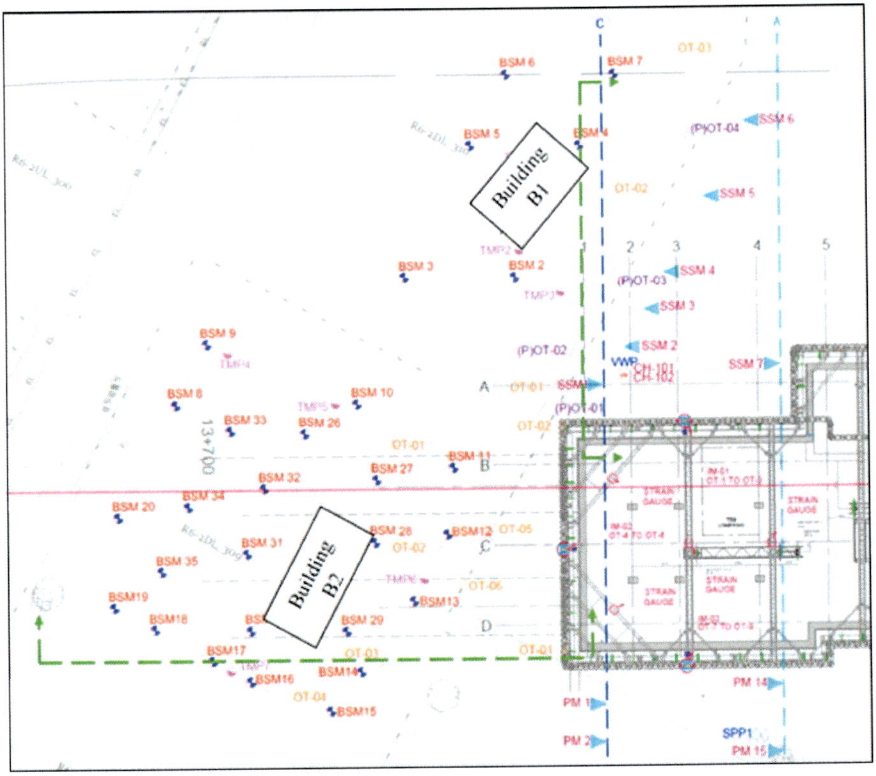

Fig. 5 Foundation layout for the buildings B1 & B2 Plan indicating cross-section considered for analysis

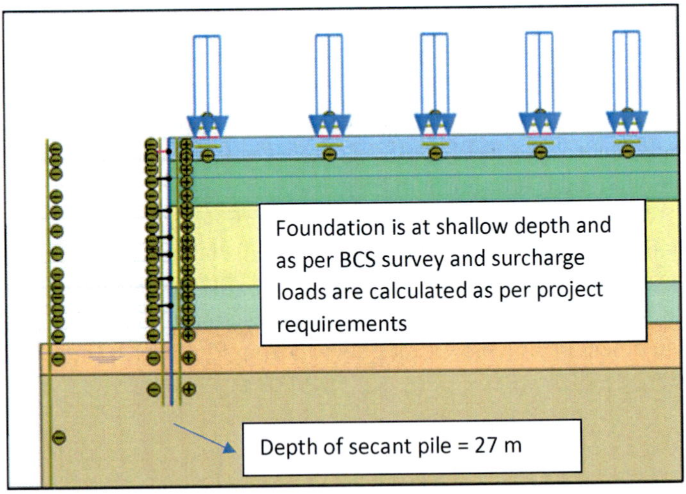

Fig. 6 Typical Excavation at Launching shafts near building B1

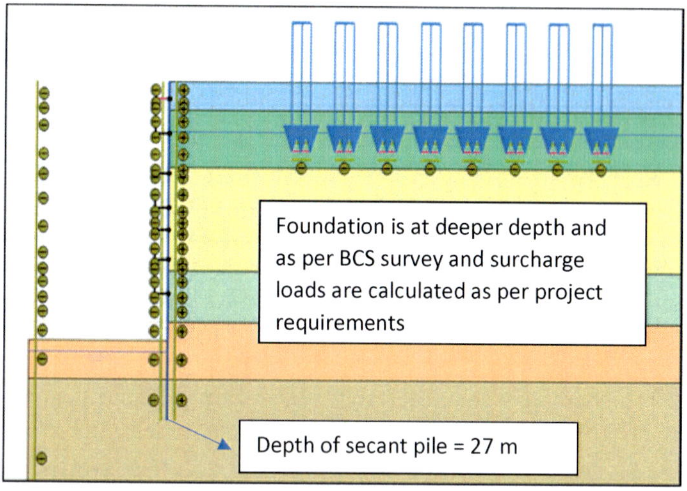

Fig. 7 Typical Excavation at Launching shafts near building B2

5 Field Instrumentation Measurements

For any underground metro construction activity, instrumentation plays a major role in confirming the assumptions made in the design and also the predicted behaviour of the support system during excavation. Therefore, instruments like inclinometers (IM) and optical targets (OT) are installed on the secant piles at the launching shaft location which is already shown in Fig. 5. In a similar manner, certain instruments are installed on the ground and adjacent buildings and structures to monitor and verify the predicted behaviour. The instruments installed in site are provided with trigger levels to check the actual measured values. Mostly building settlement markers (BSM), tilt meter plate (TMP) and optical targets for buildings are used in monitoring the adjacent buildings whereas soil settlement markers (SSM) are placed on the ground and pavement markers (PM) are kept directly on the pavements. The detailed instrumentation plan near the launching shaft location is presented in Fig. 5.

Only limited number of instruments are considered for comparing the results with analytical and numerical methods as mentioned in Table 4.

The comparison between numerical, analytical results and site-measured data explained in detail in the next section.

Table 4 Data observed from Instruments located at site used for interpretation

Instruments type	Instruments considered
Building settlement markers (BSM)	BSM 2, BSM 4, BSM 7, BSM 11, BSM 12, BSM 13, BSM 14, BSM 15
Inclinometers (IM)	IM−02
Tilt meter plate (TMP)	TMP 3, TMP 6

6 Results and Discussion

The movements of secant pile wall deflection installed at site is recorded with the help of two inclinometers IM-01 and IM-02. The construction adopted in site is exactly simulated in PLAXIS 2D and the behaviour of the secant piles is obtained. The results of the instrumented data and numerical analysis is compared and shown in Fig. 8.

From Fig. 8, it is clear that the predicted maximum deflection of secant pile is higher than the observed data which is actually recorded at site up to the final excavation level. The deflection recorded at site shows a similar trend to numerical results below 15 meters. But from the ground level to a depth of 15 m, the deflection observed in the site is higher than numerical results. The maximum deflection predicted from numerical analysis is 66 mm. However, the maximum deflection observed at site is 44 mm i.e., the maximum deflection predicted from numerical analysis is 33% higher than the actual observed value. This could be due to the support provided by the adjacent and opposite secant pile walls as the stress will be acting in three-dimensional directions. And also might be due to the water pressure on the retaining side of the secant pile wall. Hence, the deflection on the wall is comparatively higher in numerical analysis than observed instrumented data. The comparison of ground movement behind the wall obtained from numerical and analytical methods is shown in Fig. 9.

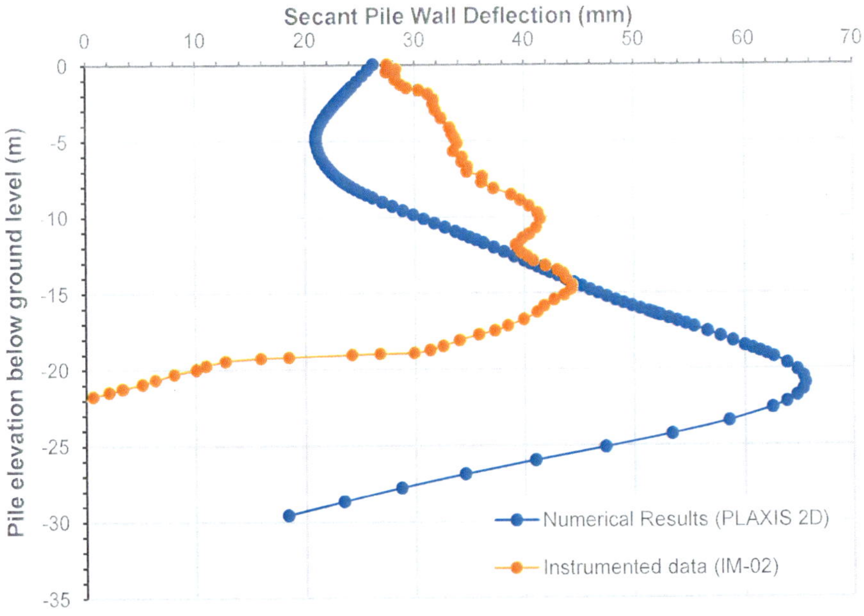

Fig. 8 Comparison of Deflection Profile of Secant pile wall

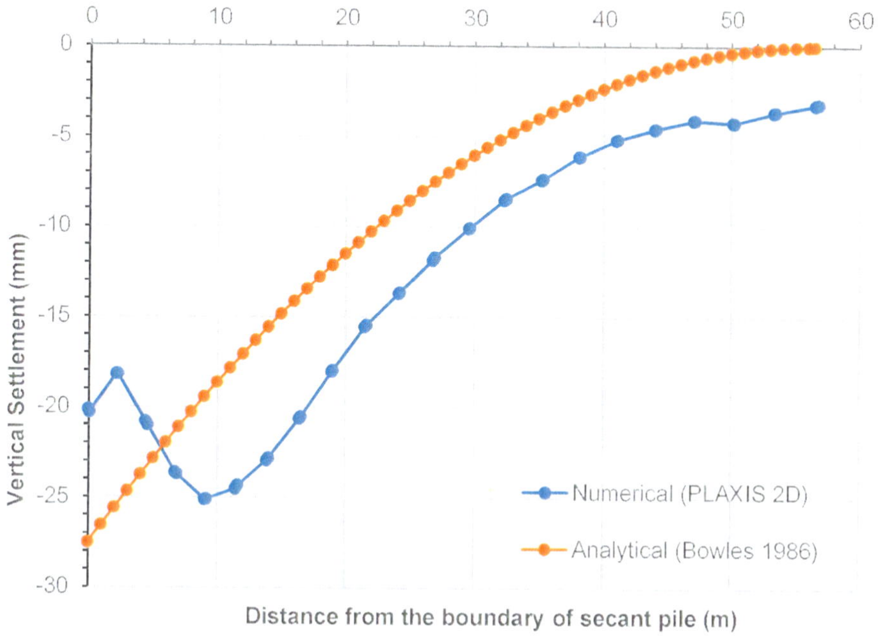

Fig. 9 Comparison of ground movement behind the secant pile wall

The results show that the maximum ground settlement predicted from analytical approach is 1% higher than the numerical results. However, the ground settlement profile remains identical beyond the distance of 10m from the wall. The ground settlement predicted from numerical results is higher than analytical approach from 6m beyond the wall. The initial variation in ground settlement obtained from the numerical analysis is due to the interaction between soil and the secant pile wall and might be due to the strut installation during excavation. Similarly, the building settlement markers, installed at site observed the settlements of the building during excavation is shown in Tables 5, 6.

From Tables 5 and 6, it is clear that settlements observed at site is much lesser than settlements predicted from numerical and analytical analysis. This might be due to conservative use of soil parameters in the design. The negative values indicate that instead of settlement, heaving has occurred at some locations of the buildings. This indicates that the building is trying to rotate but the results of tilt meter plates installed at those buildings shows no rotation at all during the excavation. From this,

Table 5 Comparison of ground movements for building B1

Building	Distance from the secant pile wall	Numerical analysis	Analytical approach	Settlement markers		
				BSM 2	BSM 4	BSM 7
B1	3 m	19 mm	25 mm	1.1 mm	−1.7 mm	−2.7 mm

Table 6 Comparison of ground movements for building B2

Building	Distance from the secant pile wall	Numerical analysis	Analytical approach	Settlement markers				
				BSM 11	BSM 12	BSM 13	BSM 14	BSM 15
B1	13 m	23.5 mm	16.5 mm	3.0 mm	3.3 mm	3.3 mm	−6.9 mm	−3.3 mm

Table 7 Results of Building rotation for Building B1 and B2

Building	Numerical analysis	Analytical method	Tilt meter plate	
			TMP 3	TMP 6
B1	0.056°	0.036°	0°	–
B2	0.027°	0.025°	–	0°

a conclusion can be drawn that either tilt meter plate is insensitive to capture the movement or the people who are living there might have disturbed the settlement markers 4, 7, 14 and BSM 15. The results of the building rotation from the numerical, analytical and tilt meter are shown in Table 7.

From Table 7, it is understood that numerical analysis gives higher values than analytical method. It is due to the fact that ground settlement is predicted higher in numerical analysis hence consequently the rotation predicted will be higher in case of numerical analysis than analytical method. However, the instruments TMP 3 and TMP 6 show that there is no rotation for the building B1 and for building B2. The safety of the building is ensured in a way that excavation being done in a controlled manner and at every stage.

7 Summary

Results of the numerical analysis and instrumented data shows that the deflection of the secant pile predicted from PLAXIS 2D is 33% higher than the deflection observed during excavation. The probable reason might be the support provided by the secant piles in opposite sides as the stress around the secant piles will be acting in three-dimensional directions. This might be due to water level at site during excavation is lesser than what was considered during design stage. The higher the water pressure, the higher the active force on the wall, which in turn increases the wall deflection from numerical analysis. Similarly, when the wall deflection is high, the ground settlements will also be high. The ground settlement predicted from analytical approach is approximately 1% higher than numerical analysis. In addition, the ground settlement from PLAXIS 2D is higher than analytical approach beyond 6 m from the wall boundary. However, the actual settlement markers installed on the buildings shows minimal amount of settlement observed during excavation. This is due to the fact that construction carried is out in a controlled manner and observations are monitored throughout the excavation phase. However, some settlement markers indicate heaving which shows that buildings try to rotate. Even the results from PLAXIS 2D and analytical approach indicates rotation may happen. However, the tilt meters installed on the buildings shows no rotation. This confirms that the excavation is carried out in a safe and progressive manner. The assumptions considered during design stage are conservative and the results predicted from numerical and analytical results are much higher than the instruments recorded at site.

References

Boscardin MD, Cording EJ (1989) Building response to excavation-induced settlement. J Geotech Eng, ASCE 115(1):1–15

Bowles JE (1990) Foundation analysis and design, 4th Edn. McGraw-Hill book company, New York, USA

Burland JB, Wroth CP (1974) Settlement of buildings and associated damage. In: Proceedings of Conference on settlement of structures, Pentech Press, London, England, pp. 611-654

Burland JB, Broms BB, de Mello VFB (1977) Behaviour of foundations and structures. In: Proceedings of the 9th International conference on Soil Mechanics and Foundation Engineering, Vol. 2. Tokyo, Japan, pp. 495-546

CIRIA C760 (2017) Guidance on embedded retaining wall design

Clough GW, O'Rourke TD (1990) Construction induced movements of in-situ walls. In: Specialty conference on Design and Performance of Earth Retaining structures, vol. 25. ASCE Special publication, pp. 439-470

Guglielmetti V. et al. (2007) Mechanized tunnelling in urban areas. Taylor & Francis Group, London, UK

Ou CY (2006) Deep excavation theory and practice. Taylor & Francis Group, London, UK

Peck, RB (1990) Deep excavations and tunneling in soft ground. In: Proceedings of the 7th International conference on Soil Mechanics and Foundation Engineering, Mexico City, State-of-the-art Volume, pp. 225-290

Numerical Study on Uplift Capacity of Helical Pile Embedded in Clay

V. Karthick Kumar and M. Muttharam

Abstract Helical piles foundation systems have been extensively used in engineering applications. It is commonly adopted to resist the compression, uplift force, overturning moment and lateral forces. Helical pile system is extensively used in many structures such as Transmission towers, Wind farms and other offshore structures because it is easy and quick to install. Realizing the importance of helical pile, an attempt is made in the present investigation to bring out the effect of spacing between the helix, blade thickness, and shaft thickness through numerical analysis. To validate the finite element model experimental investigation was carried out. In this investigation model steel piles with dual helical plate embedded in clay bed with the consistency of 0.4 was tested to study the effect of uplift loading on the behaviour of piles in tensions. Model helical pile with 12.5 mm diameter circular shaft and helix diameter of 25 mm. The model pile is embedded in clay bed for a depth of 400 mm. Parametric analysis of helical pile by varying blade thickness, shaft thickness, and the spacing of helix was carried out by using Finite Element tool Plaxis 3D. From this analysis, an increase in thickness from 4 to 10 mm increase the uplift load. Similarly, an increase in spacing of helices from 0.5 to 1.5 m increases the uplift capacity beyond that it decreases the uplift capacity. By keeping the contact area of the helix as constant as the number of a helix and helical spacing was varied, the critical number of the helix was found as 2 and the critical spacing of helix was found as 1.5 m for this study.

Keywords Critical spacing · Finite element method · Helical pile · Uplift capacity

1 Introduction

The helical pile foundation system is commonly adapted to resist compression force, uplift force, overturning moment and lateral force. When the weight of the superstructure is less than the uplift pressure acting on the structure, helical piles are

V. K. Kumar (✉) · M. Muttharam
Anna University, Tamil Nadu, India
e-mail: karthick_viswa@hotmail.com

B. Adimoolam et al. (eds.), *Deep Foundations for Infrastructure Development in India*, Lecture Notes in Civil Engineering 315, https://doi.org/10.1007/978-981-19-8598-0_16

preferred. Several critical parameters greatly influence the uplift capacity of helical pile such as the diameter of a helix, number of helices, the diameter of the shaft, the spacing between the helices, and overburden pressure. Even though the Geometry of helical pile is the same, it shows different uplift capacities in different types of soil. Ashraf et al. (1991) experimentally compared the uplift behaviour of helical screw anchor between loose sand, medium-dense sand and dense sand, it shows that failure mechanism is depending on the installation depth of the pile and angle of failure in a vertical plane does not exceed 2/3 of the angle of internal friction of sand. Ilamparuthi et al. (2002) carried out an experimental investigation on uplift behaviour of circular plate anchors embedded in loose, medium-dense, and dense dry sand. In his investigation, critical embedment ratio for uplifting capacity increases with an increase in sand density and recommended the following value of embedded ratio 4.8, 5.9, and 6.8 for loose, medium-dense, and dense sand respectively and this embedded ratio is valid only for anchors in the 100–150 mm diameter range. Nazir et al. (2015) revealed that the uplift load capacity of an enlarged base pier embedded in dry sand depends on bell angle and shaft diameter based on their experimental study. However, they also brought out that the uplift capacity decreased on increasing the bell angle from 30 to 60. Rawat and Gupta (2017) studied the pullout of helical soil nail in the sand and observed that the maximum capacity was obtained when the spacing of helix to shaft diameter ratio was less than 1.5, after which there was a small variation. Likhitha and Ramesh (2016) carried out experiments on silt deposit and concluded that uplift capacity depends on the L/d ratio, further the resistance increased with the increase in diameter of the helix. Albusoda and Abbase (2017) studied the behaviour of square helical pile of size 5 mm × 5 mm experimentally where the pile was embedded on expansive clay overlaying a sandy layer. From that experimental study, it was concluded that the uplift capacity of the helical pile was influenced by helical diameter and embedded length. Li.W et al. (2019) obtained the correlation between the uplift capacity and installation torque, which industries widely use. The uplift capacity of axially loaded piles in clays whose undrained cohesion increases linearly with depth, in that situation uplift capacity calculated with the application of axisymmetric static limit analysis approach proposed by Khatri et al. (2010) and uplift resistance has been evaluated in the form of the uplift factors, F_{cb} and F_{ct}. There are three methods for predicting pullout capacity, namely, cylindrical shear, individual bearing, and an empirical method based on installation torque. Merrifield and Smith (2010) used numerical modelling technique on plane strain multi-plate anchor foundation. The capacity of individual anchor plate within multi-plate anchor was largely independent of another anchor and depends only on the blade width. Numerical analysis of helical Pile in cohesionless soil by varying the density of sand and helical plate using Plaxis 3D by George et al. (2017). He concluded that shaft diameter does not play an important role in resisting uplift force. Embedded ratio and density play a vital role. Soil disturbance caused by the installation process is taken into account by Pérez et al. (2018). The uplift capacity was considered to be over predicted if the effect of installation was ignored. These studies show the importance of critical parameters which greatly influence the helical pile while resisting the uplift force. So it is mandatory to find out the critical spacing and the critical number of helices before going to adopt a helical pile system.

2 Experimental Test

A model helical pile of 12.5 mm shaft diameter, 25 mm helical diameter and having a length of 400 mm is used for laboratory pullout test. Failure zones along the radial direction will be 2.5 times the diameter of the helical pile and top width of the failure plane will be 4 times the diameter of the shaft. The diameter of the helix considered for experimental study in this present investigation is 25 mm and the diameter of the shaft is 12.5 mm. Influence of Stress below the pile will be up to 5 times the diameter of the helical pile recommended by Ashni and Janani (2017). So, for this experimental investigation circular tank of diameter 0.5 m and a depth of 0.6 m was selected. Properties of helical pile and model tank are shown in Table 1.

The soil used for this experiment is clay which is prepared at a consistency of 0.4, remaining properties of a clay sample used for this study is shown in Table 2.

A Clay sample was prepared at a consistency of 0.4 and fill it on a model tank. After filling the soil, model helical pile is driven up to an embedded length of 400 mm. The pullout test arrangement consists of a loading frame, model test tank prepared above clay bed and dial gauge for measuring the displacement of the model helical pile. Helical pile is subjected to tensile loading through a two-pulley arrangement

Table 1 Properties of helical pile and model tank

S.no	Components	Material type	Dimensions (mm)	Diameter (mm)	Length (mm)
1	Model tank	Steel	–	500	600
3	Hollow pile	Stainless steel 304L	1 mm thick	12.5	400
4	Helical blade	Stainless steel 304L	1 mm thick	25	–

Table 2 Properties of soil

Parameter	Value	Parameter	Value
Specific gravity G_s	2.6	IS classification (Group Symbol)	CH
Liquid limit (%)	67	Clay (%)	81
Plastic limit (%)	21	Silt (%)	16
Shrinkage limit (%)	5	Sand (%)	3
Plastic index (%)	46	Undrained cohesion (at $I_c = 0.4$) kN/m^2	13.5
Free swell index (%)	85	Natural water content (For $I_c = 0.4$)%	49

Fig. 1 Experimental setup

with a flexible wire whose one end is attached to an eye bolt in the pile model and the other end connected to the loading hanger, an experimental setup is shown in Fig. 1.

The load increment on a loading hanger is taken as an uplift force through a wire and pulley arrangement. The dial gauge of 50 mm capacity with the least count of 0.01 mm is proposed to be used to measure the displacement of the model helical pile. The load will be applied incrementally by adding weight until reaching failure. Each load increment will be maintained constant until the pile head gets stabilized. Uplift capacity was obtained from uplift load versus uplift axial displacement response is shown in Fig. 2. Reading obtained during the experiment is tabulated in Table 3.

3 Numerical Test

Further, the parametric analysis is done with the help of Finite element software called Plaxis 3D. Dimension used for the model in experimental study cannot be directly validated using Plaxis 3D software because dimensions are very small, it results in a problem in creating mesh generation itself. To overcome this error, the model helical pile is scaled to a certain factor suggested by Wood. D.M in Geotechnical Modelling.

Similitude in modelling requires selection of model dimension and stiffness so that $\frac{l}{r}\sqrt{\frac{G}{E}}$ is identical in model and prototype. So, the scale factor will be

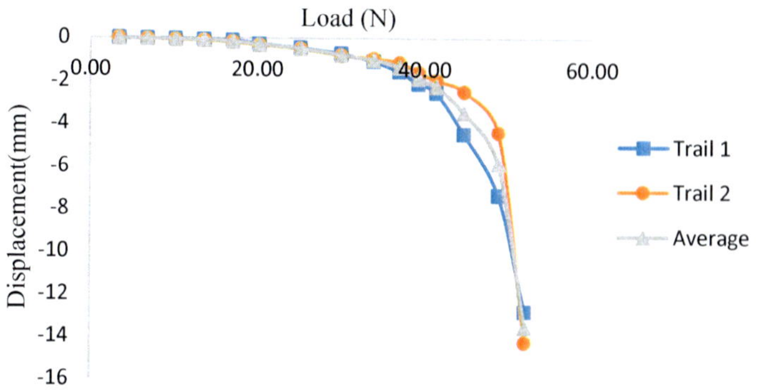

Fig. 2 Load displacement curve

Table 3 Dial gauge reading

Trail 1				Trail 2			
Load	Dial reading		Average	Load	Dial reading		Average
	Dial 1	Dial 2			Dial 1	Dial 2	
N	div	div	mm	N	div	div	mm
0	0	0	0	0.00	0	0	0
3.42	1	3	0.02	3.42	2	4	0.03
6.84	1	5	0.03	6.84	4	6	0.05
10.25	3	7	0.05	10.25	6	10	0.08
13.67	5	9	0.07	13.67	9	13	0.11
17.09	10	14	0.12	17.09	16	22	0.19
20.28	24	32	0.28	20.28	32	34	0.33
25.23	41	51	0.46	25.23	48	52	0.5
30.11	68	78	0.73	30.11	79	83	0.81
33.92	106	110	1.08	33.92	92	98	0.95
37.01	152	160	1.56	37.01	110	118	1.14
39.30	211	213	2.12	39.30	158	162	1.6
41.39	251	255	2.53	41.39	195	199	1.97
44.70	447	449	4.48	44.70	244	254	2.49
48.75	732	736	7.34	48.75	439	441	4.4
51.94	1276	1284	12.8	51.94	1426	1424	14.25

$$\frac{\left(\frac{l}{r}\sqrt{\frac{G}{E}}\right)_p}{\left(\frac{l}{r}\sqrt{\frac{G}{E}}\right)_m} = 1 \tag{1}$$

Table 4 Engineering properties of soil and steel

Parameter	Value	Unit
Soil		
Model	Mohr–coulomb undrained B	–
Unit weight	14	kN/m^3
Young's modulus	2455	kN/m^2
Poisson ratio	0.4	–
Undrained cohesion	13.5	kN/m^2
Steel		
Thickness	1	cm
Unit weight	78.5	kN/m^3
Young's modulus	1.93×10^8	kN/m^2
Poisson ratio	0.3	–

Here the dimensions are scaled up to n times and the Load is scaled up to n^3 times suggested by Wood. D.M in Geotechnical Modelling.

Engineering parameters of soil and steel used for the numerical analysis is given in Table 4.

Experimental results are scaled up and compared with numerical results. Comparison of numerical and experimental results are shown in Fig. 3. Results obtained from the numerical analysis by using fine coarseness mesh match with experimental results and the variation was less than 10%. Hence, the fine mesh will be adopted for further parametric analysis.

Parameters such as the thickness of the shaft, the thickness of the helical blade and the spacing of helices varied for the above-generated model. Load displacement curve for the varied parameter is shown respectively in Figs. 4–9.

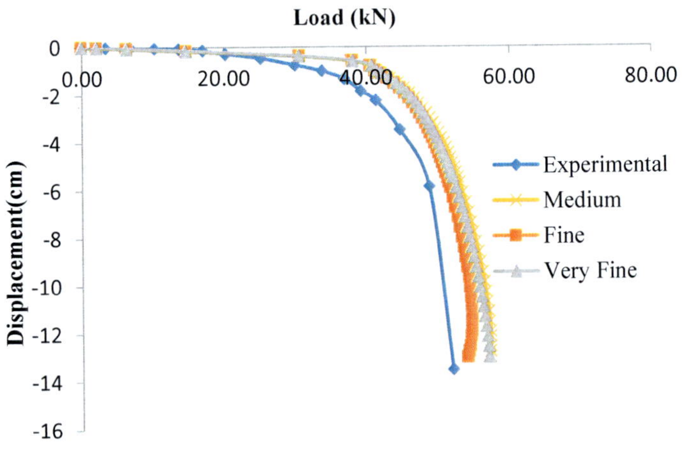

Fig. 3 Comparison of experimental and numerical results

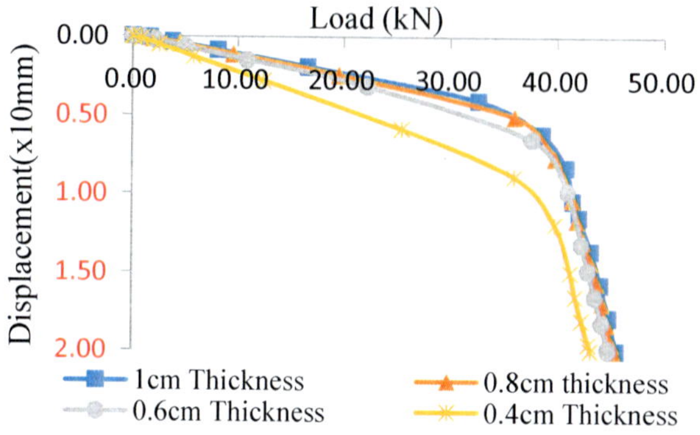

Fig. 4 0.5 m spacing of helix

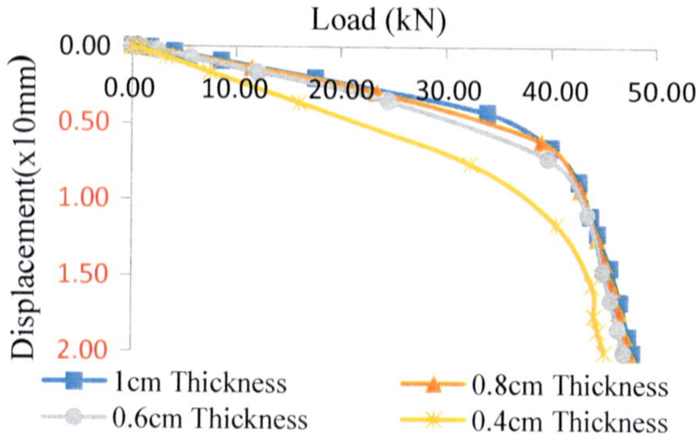

Fig. 5 1 m spacing of helix

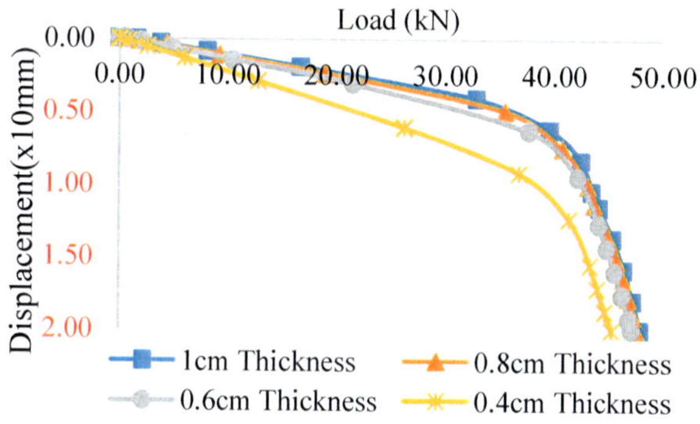

Fig. 6 1.5 m spacing of helix

After varying spacing and thickness of dual helix, the critical number of the helix for the same contact area was obtained. For example, the above case contact area of the dual helix is 73593.75mm^2. By maintaining the above contact area constant, the number of helices was varied as 1 and 3. So, the diameter of the helix for the single helical pile is 353.5 mm and for triple helix 204 mm. Result obtained from the analysis is tabulated in Table 5.

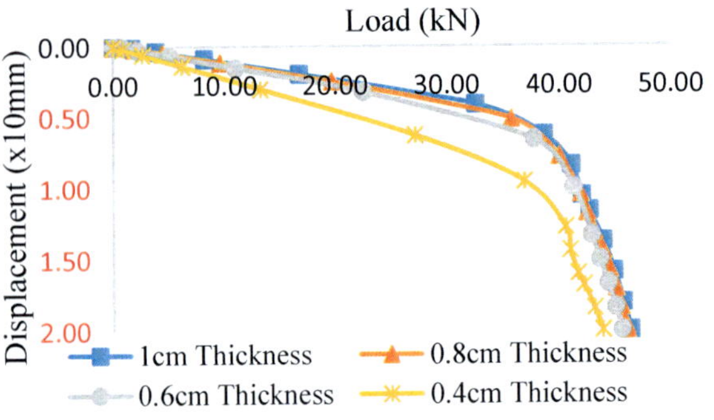

Fig. 7 2 m spacing of helix

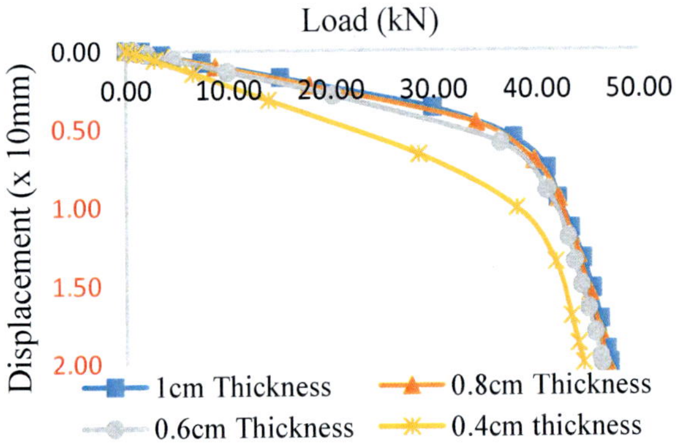

Fig. 8 2.5 m spacing of helix

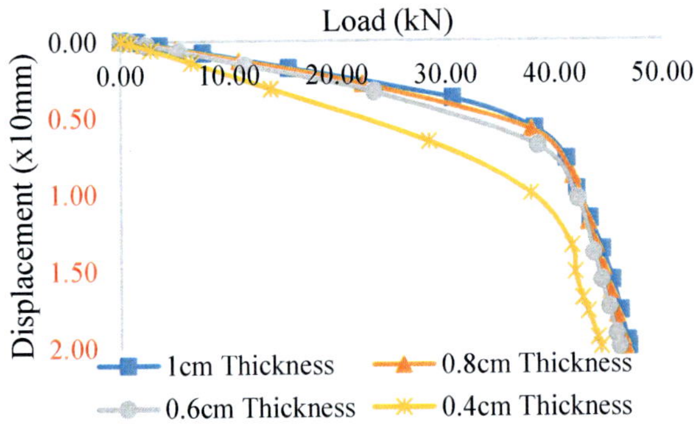

Fig. 9 3 m spacing of helix

4 Summary

From the parametric analysis, results are compared with different spacing of helices for 250 mm diameter helical pile, data obtained from the results for 250 mm diameter helical pile is tabulated in Table 6. Variation of the uplift load for the spacing of helices is shown in Fig. 10.

Table 5 Uplift load for 12 mm displacement

No of helix	Spacing (m)	Uplift capacity (kN)
Single helix	–	41.3
Double helix		
	0.5	38.37
	1	39.39
	1.5	39.88
	2	38.44
	2.5	37.32
	3	36.77
Triple helix		
	0.5	33.37
	1	31.48
	1.5	30.04

Table 6 Uplift load of different spacing for 12 mm displacement

Thickness	0.5 m Spacing	1 m Spacing	1.5 m Spacing	2 m Spacing	2.5 m Spacing	3 m Spacing
	Load	Load	Load	Load	Load	Load
mm	kN	kN	kN	kN	kN	kN
4	31.72	32.75	33.62	33.1	32.5	31.21
6	34.51	35.82	36.54	35.5	34.3	33.01
8	36.22	37.46	38.3	36.96	35.46	34.88
10	38.47	39.39	39.88	38.44	37.32	36.77

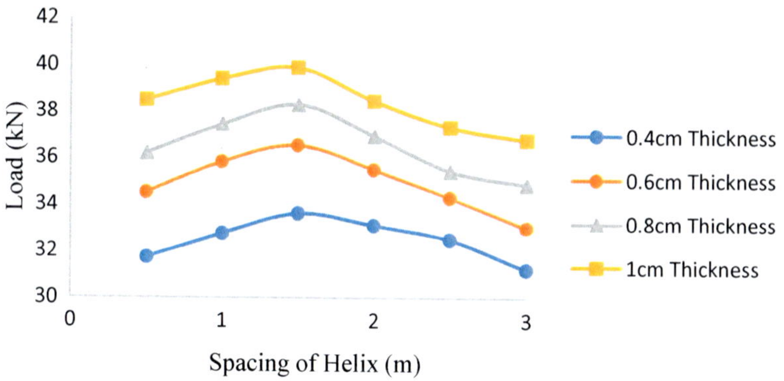

Fig.10 Variation of uplift load due to spacing

5 Conclusion

The spacing of helices shows greater variation in the uplift capacity, it decides the mode of resistance in the soil. Varying the spacing of helix from 0.5 m to 3 m in pile shows an increment in uplift capacity of up to 1.5 m after that a small decrement in the uplift capacity. This decrement is due to the transition from cylindrical shear resistance to the individual bearing. When the spacing between the helices is large, the pile will resist the uplift load by an individual bearing so the uplift capacity decreases accordingly. The spacing at which the transition occurs is called critical spacing. This critical spacing is greatly influenced by the diameter of helix. For the identical geometry helical pile, the critical spacing will vary depending upon the properties of the soil. Apart from spacing, the number of helices also influences the uplift capacity. Table.5 shows the uplift capacity of various numbers of the pile. From the results, the uplift capacity of the single helical pile and the dual helical pile at 1.5 m spacing shows almost the same resistance. Hence, the critical number of the helix for this case is two. Studies on the performance of helical pile foundation can be extended in many ways such as the uplift capacity of the helical pile under both lateral and axial load in clay, Uplift capacity of the helical pile group in both clay and sand for combined loading and Analysis of critical spacing of a helix for the combined loading in different soil properties.

Acknowledgements Authors would like to acknowledge the Division of Soil Mechanics and Foundation Engineering, Anna University, Chennai, India for the laboratory and software facilities provided to carry out this experiment.

References

Albusoda BS, Abbase HO (2017) 'Performance assessment of single and group of helical piles embedded in expansive soil'. Int J Geo-Eng 8(1)

Ashni M, Janani V (2017) Experimental study on the pull-out capacity of the helical pile in clayey soil. Int J Civ Eng Technol 8(4):1514–1521

Ghaly A, Hanna AM, Members ASCE (1991) 'Uplift behaviour of screw anchors in sand I: Dry sand'. J Geotech Eng 117(5).

George BE, Banerjee S, Gandhi SR (2017) Numerical analysis of helical piles in cohesionless soil. Int J Geotech Eng 6362:1–15

Ilamparuthi K, Dickin EA, Muthukrisnaiah K (2002) Experimental investigation of the uplift behaviour of circular plate anchors embedded in the sand. Can Geotech J 39(3):648–664

IS 2911–4: Code of practice for design and construction of pile foundations, Part 4: Load test on piles.

Khatri VN, Kumar J (2010) Uplift capacity of axially loaded piles in clays. Int J Geomech 11(1):23–28

Li W, Deng L (2019) Axial load tests and numerical modelling of single-helix piles in cohesive and cohesionless soils. Acta Geotech 14(2):461–475

Likhitha H, Ramesh BR (2016) 'The Experimental Study on Uplift Bearing Capacity of Helical Piles in Silt Deposits', pp. 9363–9369.

Merrifield RS, Smith CC (2010) The ultimate uplift capacity of multi-plate strip anchors in undrained clay. Comput Geotech 37(4):504–514
Nazir R, Moayedi H, Pratikso A, Mosallanezhad M (2015) The uplift load capacity of an enlarged base pier embedded in dry sand. Arab J Geosci 8(9):7285–7296
Pérez ZA, Schiavon JA, de Tsuha CHC, Dias D, Thorel L (2018) Numerical and experimental study on the influence of installation effects on the behaviour of helical anchors in very dense sand. Can Geotech J 55(8):1067-1080
Rawat S, Gupta AK (2017) Numerical modelling of a pullout of helical soil nail. J Rock Mech Geotech Eng 9(4):648–658
Wood DM (2017) Geotechnical modelling. Appl Geotech 1:1–497

Safe and Efficient Geo-Construction

Rectification Measures and Restoration of Distressed Pump Foundations

Sampat Raj, Geethanjali Koppolu, and V K Panwar

Abstract At a Fertilizer plant located in the southern part of India, 74 numbers of pump foundations have shown non-uniform settlements causing the tilting of pump foundations under self-weight primarily due to the presence of varying thickness of the heterogeneous subsoil beneath the pump foundations and soil erosion because of water seepage through deeper excavations in the close vicinity. Various corrective measures have been explored and taking due cognizance of the prevailing subsoil conditions, site constraints, and constructability aspects, micropiling has been adopted. This paper presents the site problem and details of remedial measures adopted for rehabilitation of the distressed pump foundations. The main objective of this paper is to provide guidelines for designer during design and detail engineering stage to avoid similar nature of problems and also to provide a solution concept to tackle similar nature of problem(s) if encountered at the site.

Keywords Pump foundation · Settlement · Micropile

1 Introduction

A fertilizer plant comprises primarily of two major units namely ammonia and urea units to cater the process requirements. These units house various facilities viz. pipe rack, technological structures and many other critical equipments such as process columns, vessels, tanks, pumps etc. which are generally located in a close vicinity owing to their functional and process requirements.

The foundation system for the above structures/facilities is firmed up based on the site, subsoil conditions and functional/loading requirements. In the case of shallow

S. Raj (✉) · G. Koppolu · V. K. Panwar
Engineers India Limited, New Delhi, India
e-mail: sampat.raj@eil.co.in

G. Koppolu
e-mail: geethanjali.k@eil.co.in

V. K. Panwar
e-mail: vk.panwar@eil.co.in

B. Adimoolam et al. (eds.), *Deep Foundations for Infrastructure Development in India*, Lecture Notes in Civil Engineering 315, https://doi.org/10.1007/978-981-19-8598-0_17

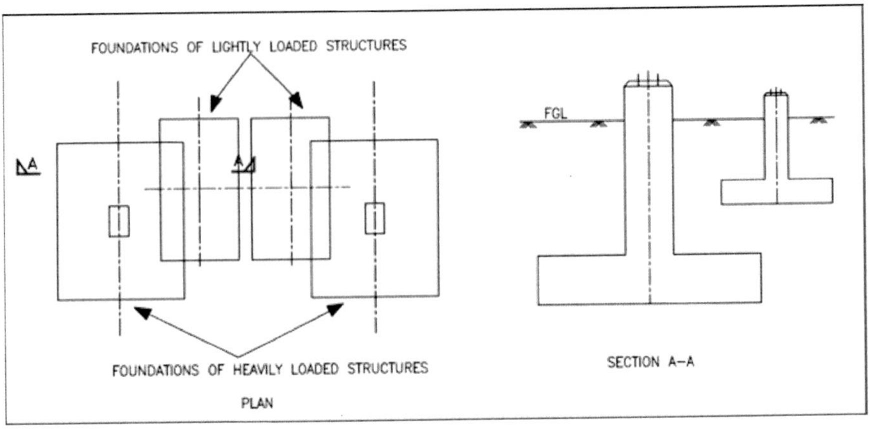

Fig. 1 Schematic view of heavy and light equipment foundations in close proximity

foundation, it is a common practice that the foundations of heavily loaded structures rest in naturally competent strata whereas, many times foundations of lightly loaded structures rest at shallower depths. The typical foundation layout of equipment in the process unit is shown in Fig. 1 indicating the proximity and difference in the level of heavier and lighter equipment foundations. Deeper foundations are always constructed in the initial stage. After the construction of the deeper foundations, backfilling is carried out by putting the excavated/suitable borrowed material in the foundation pits. Lightly loaded equipment foundations located adjacent to the deeper foundations may rest on backfilled soil, completely or partly depending on the footprint area of footings.

At a fertilizer plant in the southern part of India, around 74 numbers of pump foundations housed in ammonia and urea units exhibited tilt and non-uniform settlements. These pump foundations were located adjoining to the relatively deeper pipe rack foundations. Pumps are the "Heart" of any industrial plant and are treated as critical machinery. Pumps are generally supported on block foundation and structural integrity of foundation system is of paramount importance for running the plant safely and with optimum efficiency. Any damage/differential settlement of the pump foundation shall adversely affect the performance and life of the equipment due to increased vibrations, misalignments within the equipment and also with the battery limit interfaces. This paper presents the site problem and details of remedial measures adopted for rehabilitation of the distressed pump foundations.

2 Topography of Site

In the past, a smaller capacity coal based fertilizer plant was operational at the proposed site location, which was completely dismantled to construct the proposed higher capacity gas based fertilizer plant. Deposits and heaps of fly ash were found

Fig. 2 Fly ash inside plant boundary

at various locations inside the plant boundary. Figure 2 shows fly ash inside the plant boundary. The original ground levels in ammonia and urea units were varying from RL 165.0 to RL 170.0. Whereas, the finished ground level (FGL) is RL 167.0. Fly ash was encountered in the urea unit which was completely removed and replaced with suitable soil during site grading works.

3 Subsoil Data

A detailed subsoil exploration program consisting of boreholes, static cone penetration tests, crosshole tests and laboratory testing was carried out during the initial stage of plant set-up to characterize the subsoil and to evaluate the geotechnical design parameters. Based on the boreholes it is revealed that top soil up to 4.50 m depth in ammonia unit is very stiff to hard clay underlain by highly weathered to moderately weathered rock extending up to the termination depth of boreholes (i.e. 20 m).

However, in the urea unit top soil, up to 3m thick is fill material which was laid during general site grading work as a replacement for fly ash. The soil profile below this layer is more or less uniform in both the units. The typical soil profiles of ammonia and urea units are shown in Fig. 3. The ground water table was encountered at 3 to 7 m depth below existing ground level during the period of investigation.

4 Problem Description

Based on the subsurface exploration data, it is apparent that the subsoil is well suited for supporting the foundations at shallower depths. A schematic view of the foundation system is shown in Fig. 4 indicating the proximity and level difference

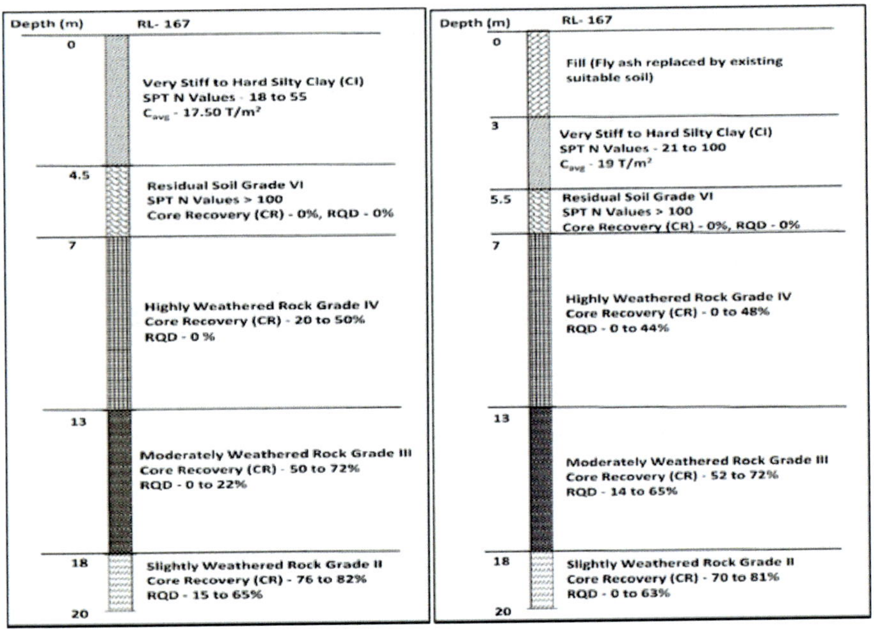

Fig. 3 Typical soil profiles of ammonia and urea units

of pipe rack and pump foundations. Pumps were located adjoining to the pipe rack structure whose foundation depth is 3.50 m below the finished ground level (FGL). Block type pump foundations having widths varying from 1.0 to 2.70 m and lengths varying from 1.75 to 7.10 m were placed at shallower depth over improved ground (i.e. 500 mm thick compacted fill). The weight of the pumps varied from 0.80 to 16.20 t and the foundation weight varied from 5 to 60 t. Majority of the pump foundations were placed at 1.0 m depth below FGL. During execution, pipe rack foundations were constructed first which required excavation up to 3.50 m depth below FGL followed by construction of the foundation/pedestals and backfilling of the foundation pit. The specified compaction criteria for the backfilling of foundation pit was that the excavated soil should be laid in layers of maximum 200 mm loose thickness and compacted to 95% of modified proctor density as per IS 2720 (Part 8) (1983). Since pumps were located so close to the pipe rack, the pump foundations were partially coming on the pipe rack foundation which resulted in variable thickness of backfilled soil below the base of pump foundations.

During the construction, the site experienced continuous heavy rain for a period of 3 days and post rains the pump foundations have shown non-uniform settlements causing the tilting of foundations under self-weight as shown in Fig. 5. Settlements were measured at all four corners of pump foundations and the observed settlements were higher than the acceptable limits. The maximum differential settlement experienced was about 5 to 348 mm for various pump foundations. A detailed analysis was

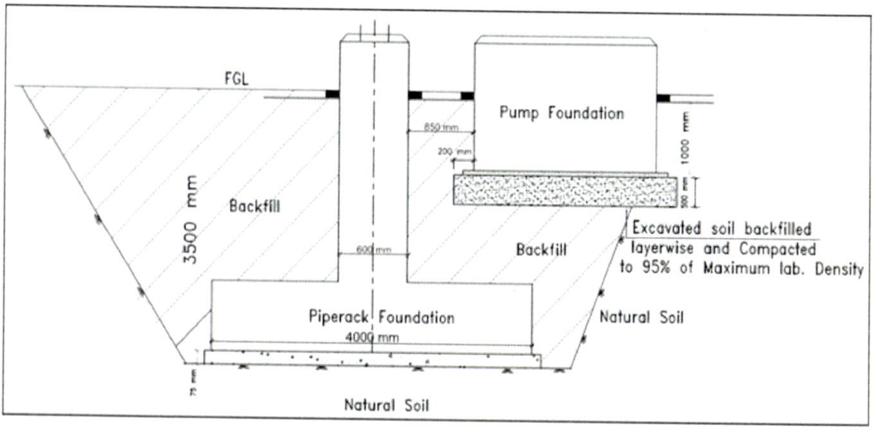

Fig. 4 Typical arrangement of pipe rack and pump foundations

carried out to determine the root cause and to suggest the remedial measures for rectification of the distressed foundations. Inadequate compaction and presence of soils having varying stiffness below the pump foundations were found to be the primary reasons for tilting of the foundations. In addition, a parallel activity of laying an underground piping corridor adjacent to the pump foundations was underway, resulting in erosion of the compacted soil around/below the foundations during heavy rains. As construction sequence is the most important aspect at any site, it is apparent that a clear deviation to the established practice has been taken at the referenced site.

Fig. 5 Differential settlement of pump foundations under self-weight

5 Remedial Measures

Few pump foundations were selected for preloading up to 2.50 times the pump weight and settlement readings were measured for 72 h to determine the potential of further settlement. Figure 6 depicts the time-settlement curve for one of the pump foundations, namely the "ammonia solution pump" having a pump weight of 9.50 t and a preload weight of 23.15 t. It was observed that the settlement was continuous, implying that the pump foundations may undergo further settlement with time. The measured settlement at all four corners of the pump foundation was also varying. The settlement at corners A and C was less than that at corners B and D, indicating the tilting of the pump foundation.

The following remedial measures were explored considering subsoil conditions, site constraints, techno-economical and constructability aspects.

- Lifting of pump foundations and placing it back on a bed of lean concrete extending up to natural ground level.
- Lifting of pump foundations and placing it back on pile foundation.
- Dismantling of pump foundations and casting new foundations on bed of lean concrete extending up to natural ground level.
- Dismantling of pump foundations and casting new foundations on pile foundation.
- Providing micropiles around the perimeter of pumps and integrating it with pump foundation by concrete skirt wall.
- Chemical grouting.

Based on the critical review of all the above options, it was decided to provide micropiles around the perimeter of pumps and integrate both by concrete skirt wall. The remedial measures were intended to arrest further settlement and transfer the load to the competent substrata.

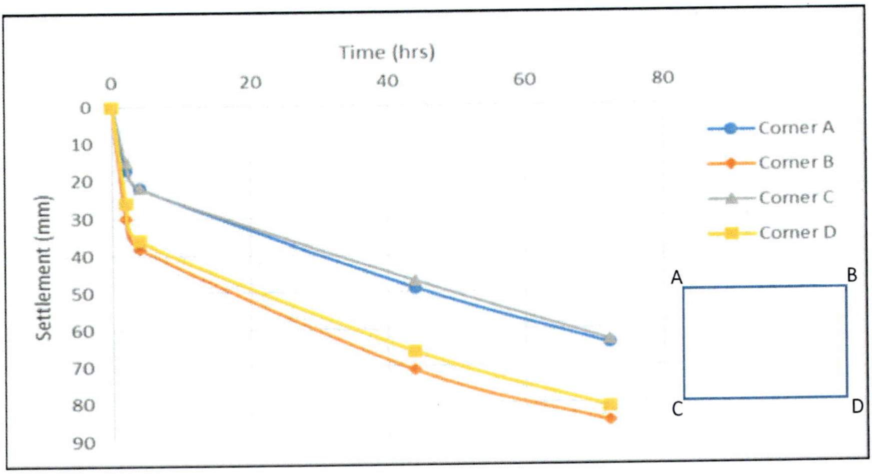

Fig. 6 Time-settlement curve

6 Design Considerations

A micropile is a small diameter (typically less than 300 mm), drilled and grouted non-displacement pile that is typically reinforced. A micropile is constructed by drilling a borehole followed by placing steel reinforcement and grouting the hole. Micropiles are proven techniques to underpin existing structures.

Micropiles were designed as per the provisions of FHWA NHI-05-039 (2005). The diameter of the micropiles was 200 mm. Design parameters used for the evaluation of structural and geotechnical capacity of the micropile are shown in Table 1.

Micropiles were designed duly considering all the technical aspects viz. ultimate bond stress between reinforcement and grout, ultimate bond stress at grout and surrounding soil interface. Axial capacity of the micropile was calculated by considering only the skin friction component. A working stress approach was followed for micropile capacity calculation and factor of safety (FOS) was applied as 2.5.

Grout to ground ultimate bond strength (α_{bond}) for the clay is 50–120 kPa as per FHWA (2005). However, for the design purpose grout to ground ultimate bond strength was considered as 75 kPa. Allowable geotechnical bond capacity of micropile was calculated as 70 kN. However, for the design purpose it was considered as 50 kN.

A steel deformed bar (Fe 500 D) of 25 mm diameter was placed centrally. Micropiles were terminated into hard clay/weathered rock strata with minimum embedment of 3 m. However, at locations where pump and pipe rack foundations were overlapping, micropiles were terminated on the top of the pipe rack foundations.

Micropiles were installed around the pump perimeter and connected to the distressed pump foundations by constructing skirt walls all around. The concrete surface of the pump foundations was connected to the skirt wall using RE500V3 adhesive anchoring system. Figure 7 shows the typical arrangement of micropiles with the existing pump foundation system.

Table 1 Design parameters

Parameters	Value
Unconfined compressive strength of grout (f_c)	30 N/mm^2
Yield stress of steel (Fy-bar)	500 N/mm^2
Diameter of micropile	200 mm
Diameter of rebar	25 mm
Grout to ground ultimate bond strength (α_{bond})	75 kPa
Bond length (L_b)	3 m
FOS	2.5

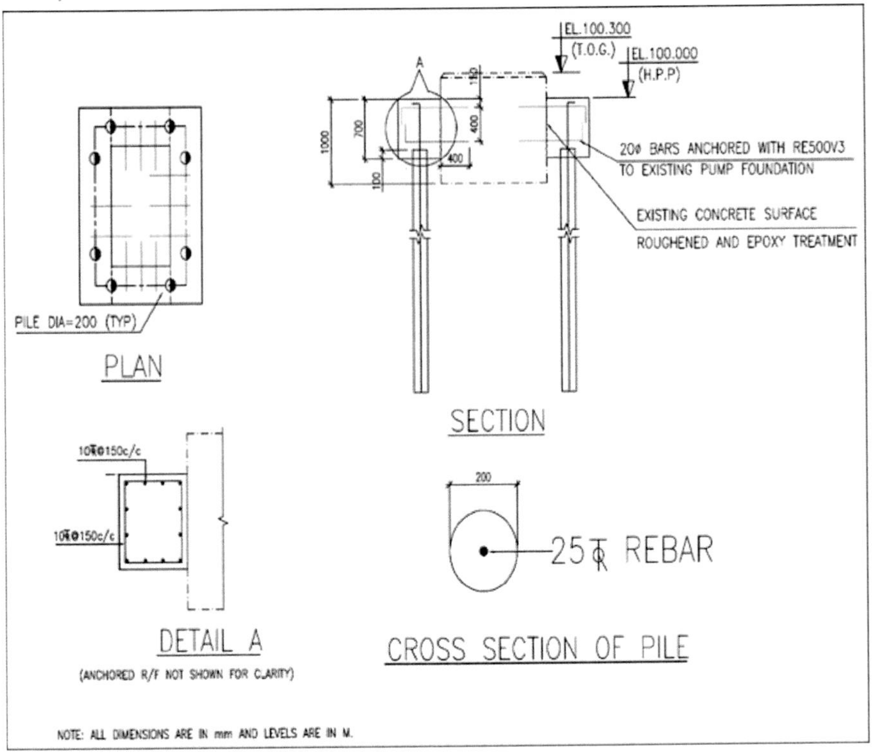

Fig. 7 Typical arrangement of micropiles with existing pump foundation system

7 Post Improvement Performance And Concluding Remarks

The remedial measures were executed successfully after overcoming all impediments. Post treatment, periodic settlement measurements were taken at site for all the treated foundations and no further settlements were observed. Post treatment performance of the pump foundations has been found satisfactory.

This case study demonstrates the importance of construction sequence along with proper selection of suitable structural fill material and their compaction criteria. All activities which require deeper excavations such as laying of underground piping corridors, drainage network etc. shall be executed in initial stages.

The type of backfill material should be such that proper compaction can be achieved as per the design requirements. The backfilled material below the lightly loaded structures should be compacted thoroughly. The compaction criteria shall be religiously followed and the suitable machinery/compactor/roller shall be deployed to achieve the design compaction criteria.

Acknowledgements The authors express their gratitude to the management of Engineers India Limited for granting permission to publish this paper.

References

FHWA NHI-05-039 (2005) Micropile design and construction.
IS: 2720 (Part 8)-1983, Reaffirmed 1995. Method of test for soils: Determination of water content – dry density relation using heavy compaction.